OBERGANSLBACH

Here Am I–
Where Are You?

Also by Konrad Lorenz

King Solomon's Ring

Man Meets Dog

On Aggression

Studies in Animal and Human Behavior

Civilized Man's Eight Deadly Sins

Behind the Mirror: A Search for a Natural History of
Human Knowledge

The Year of the Greylag Goose

Konrad Lorenz in Grünau.

Konrad Lorenz

Here Am I–
Where Are You?

The Behavior of
the Greylag Goose

*In collaboration with Michael Martys
and Angelika Tipler*

TRANSLATED BY ROBERT D. MARTIN

A Helen and Kurt Wolff Book
Harcourt Brace Jovanovich, Publishers
New York   San Diego   London

HBJ

Copyright © 1988 by R. Piper GmbH & Co. KG, München
English translation copyright © 1991 by
Harcourt Brace Jovanovich, Inc.

Library of Congress Cataloging-in-Publication Data
Lorenz, Konrad, 1903–1989
[Hier bin ich—wo bist du? English]
Here am I—where are you?: the behavior of the greylag goose/
Konrad Lorenz, in collaboration with Michael Martys and Angelika
Tipler; translated by Robert D. Martin.—1st U.S. ed.
p. cm.
Translation of: Hier bin ich—wo bist du?
"A Helen and Kurt Wolff book."
Includes index.
ISBN 0-15-140056-3
1. Greylag goose—Behavior. I. Martys, Michael. II. Tipler,
Angelika. III. Title.
QL696.A52H6813 1991
598.4'1—dc20 90-23871

Designed by G.B.D. Smith
Printed in the United States of America
First United States edition
A B C D E

To the memory of Oskar Heinroth
this book is dedicated.

Contents

Preface

This book has been many years in the preparation—in fact, the better part of my life. And yet that long preparation has not led to any conclusive results. This is always true of scientific research. Each problem we are lucky enough to solve generates, at the very moment of solution, new problems. Although I can say without immodesty that this book represents the most complete investigation to date of the ethology of a higher organism and its social system, that certainly does not mean that we have finished our analysis. To the contrary, what has been achieved so far is of value only if it serves as a tool for further research. The descriptive study of an animal species always focuses on the individual, on the phenotype of the single animal. But an exact knowledge of everything a greylag goose can do provides no more than the starting point from which to gain an understanding of the interactions between individuals in the larger system of the flock. But when, in fact, can we refer to a flock as a social system? At what lower limit is a flock no longer a flock but merely a clump of geese?

It is at this point that the question of the survival value of individual behavioral systems arises. We ethologists are often criticized—especially by sociobiologists—for assuming that any form of behavior with an obvious function (such as jealous responses or the struggle for higher rank) plays an important role in the survival of the species. This criticism is only partially justified. Ethologists have long recognized—going back to Darwin—that there are some structures and behavior patterns that aid the reproductive success of the individual without benefiting the species as a whole. The phenomenon of intraspecific selection, which brings no improvement in the level of adaptation of the species as a whole, is a familiar problem. In Socratic style, Heinroth used to say: "The working habits in civilized society are second only to the secondary feathers of the Argus pheasant in the scale of useless by-products of natural selection."

Human beings, more than any other social animal, are menaced by the deleterious effects of intraspecific selection. It is therefore no waste of our time to study intraspecific selection in another species, where it appears to proceed in a comparable fashion. The longitudinal investigation of greylag geese deals with questions that concern humans as well.

I single out two of my coworkers in Grünau, Angelika Tipler-Schlager and Michael Martys, for special thanks, in recognition of the fact that they have devoted their professional lives to the solution of these very questions.

Acknowledgments

I am greatly indebted to the Max Planck Society for the Advancement of Science for appreciating that longitudinal research on the social life of geese can yield fruit only if it is conducted over an extended period of time. It was in this spirit that the society agreed to continue its funding of the project for several years after my retirement in 1973 as director of the Max Planck Institute for Behavioral Physiology, in Seewiesen. I am also grateful to the Austrian Ministry for Science and Research and the Austrian Academy of Sciences for taking over the funding. Invaluable help also came from the Duke of Cumberland and the Cumberland Foundation, under the presidency of Chief Forestry Officer Karl Hüthmayr. I express my deepest gratitude to all.

This book, more than most, is the product of a cooperative undertaking. The collection of the basic material began many years ago, as did the long line of coworkers who have helped me. The foundation was laid by my nurse, Resi Führinger, who taught me how to rear domestic ducks and other precocial birds even before I started school. I learned a great deal from her about the duck family and about greylag geese.

Many of my assistants have acted as parents to flocks of young geese—an activity I strongly recommend for any student of ethology. Several of these gosling caretakers are now established ethologists in their own right, some of them university professors. I learned from every one of them, although I cannot mention them all by name. Those who later published their observations and are cited in my bibliography played a particularly important role in our accumulation of data.

One of the earliest observers of the geese was Helga Mamblona-Fischer. Her analysis of the motivational basis of the triumph ceremony is still valid. Sybille Kalas-Schäfer discovered the hierarchy within the family group, whose existence Oskar Heinroth had denied in discussions with me. From Brigitte Dittami-Kirchmayer I learned the importance of avoiding stress situations when hand-rearing goslings. She conducted an apparently harmless experiment that consisted simply of withholding greeting from a particular gosling. Its dramatic results struck us as so cruel that no member of our team has ever been willing to repeat it.

I also wish to thank all the technicians who did the detailed recording of the geese's behavior, with special mention to Heidi Buhrow and Gudrun Lamprecht-Bracht.

FROM THE TRANSLATOR

The invitation to translate this last book by Konrad Lorenz has provided me with a welcome opportunity to pay a personal tribute. Konrad had a major influence on me when I was a postgraduate student, leading me to appreciate the value of comparative study of the behavior of living organisms for the appreciation of evolutionary relationships. I offer this translation as a token of my respect for Konrad as a scientist and in recognition of his much-valued friendship.

I would also like to thank another graduate of the Lorenz school, Gustl Anzenberger, who not only provided advice on the translation of various difficult terms but also checked the entire manuscript.

ROBERT D. MARTIN
Professor and Director
Anthropologisches Institut und Museum
Universität Zürich-Irchel

Here Am I–
Where Are You?

Introduction

Why the Greylag Goose?

I am often asked why I chose to concentrate so much of my research on the greylag goose. There are a number of reasons, most of which are explained in detail in the later chapters on methods and the use of analogies. It is, indeed, interesting as well as revealing to probe the origins of the lifetime pursuit of any investigator. To identify the main factor that led him to chose his subject. The investigator's personal history certainly exerts a strong influence, and I cannot claim that rational considerations alone determined the course of my studies.

When I was a child, I wanted to be an owl, because owls did not have to go to bed at night. But just at that time something quite wonderful happened. My bedtime reading consisted of stories from a book by the Swedish author Selma Lagerlöf, *Wunderbare Reise des kleinen Nils Holgersson mit den Wildgänsen* (Little Nils Holgersson's Wonderful Journey with the Wild Geese). It made me see that owls had a great handicap. They could not swim or dive, things I had recently learned to do, and I decided right then that I wanted to join the ranks of the waterfowl instead. When it eventually dawned on me that I could not actually *become* an aquatic bird, I was determined that I would at least *own* one. Thanks to Selma Lagerlöf, it was a goose I wanted. But my mother, who was rightly worried about the fate of the flowers in her garden, was not willing to put up with a goose. Luckily, I soon found a solution. Our neighbor had a clutch of domestic ducklings being led around by a clucking hen. My mother finally yielded to my begging and bought me one of the ducklings, although my father was against it. He felt that trusting a freshly hatched duckling to a six-year-old boy amounted to cruelty to animals, and he did not expect the duckling to live very long. But in this particular prognosis the great doctor turned out to be wrong. My Pipsa survived for some fifteen years, close to the maximum for a domestic duck.

Even in those early days, my future wife and I had some interests in common, and she was given a duckling from the same clutch a day later. When I look back now and remember how much we learned from our two ducks, they almost seem to be my most influential teachers. We took it for granted from the beginning that the ducklings would direct toward us the behavior patterns they normally displayed toward their natural mother. We were not in the least surprised to have them follow us everywhere.

I can still remember, as if it were yesterday, squatting on the flagstones of our huge kitchen in Altenberg with the first duckling, just after we bought it. It was standing in front of me with its neck stretched upward and crying in a series of single-syllable distress calls (Figure 1). Conscience-stricken because it was at my bidding that the duckling had been taken from its mother, I tried to comfort it by imitating the special call a mother duck gives to summon her ducklings. The duckling broke off crying and uttered a two-syllable contact call—or, as Oskar Heinroth has dubbed it, the conversation call. At that, I backed away on all fours and quacked more actively, upon which the duckling waddled up to me. As it drew near, it uttered the contact call with increasing frequency, and I replied in kind. The displays and calls of the domestic duck are no different from those of its wild relative the mallard, and among adult ducks this so-called *reb-reb* palaver is the typical form of greeting. It is functionally analogous—and probably homologous in an evolutionary sense—with the

Figure 1: *A duckling calling in distress.*

greeting ceremony and the contact call of the greylag goose. (The terms analogous and homologous are discussed in the next chapter.)

Having two children of different ages acquire two ducklings on successive days led to an unplanned experiment on the phenomenon we now know as imprinting, or the fixation of an innate drive on a particular object. In the mallard, the process of imprinting to follow is limited to a period of a few hours in the duck's development and it is precisely this limitation to a specific phase of development that distinguishes imprinting from other forms of learning. My duckling had just left the nest and was more strongly bonded to me, and more obedient in following, than was the duckling that belonged to my future wife—though she always denied this afterward.

We failed to notice something else that happened, however. I, a six-year-old boy, became imprinted on birds of the family Anatidae as the main interest of my life, whereas my nine-year-old future wife proved immune to this mild form of insanity. My love of anatids, which possessed me then and remains with me today, perhaps shows that irreversible imprinting can also occur in human beings.

Although, in that summer of 1909, we felt that we had outgrown the game of pretending to be ducks, we accepted our roles as duck mothers with passion and devotion. We waded along the flat banks of the Danube seeking out pools rich in insect life and joyfully watched our ducklings gobble up this natural food. We learned to recognize when one of our foster children produced a distress call (lost piping) and to respond in the appropriate way when one of them was cold or hungry. We could soon distinguish the good-taste call the ducklings produced when they discovered tasty insect prey in the mud (especially caddis-fly larvae). And whenever we heard the trilling that ducklings and chicks utter when they need to be warmed before going to sleep, we would find a suitable fold or pocket in our clothing and warm our charges with our bodies. We spent the entire summer looking after them in this way.

My love for the mallard soon spread to other members of the family Anatidae, and by the time I was in secondary school I knew quite a lot about the different behavior patterns of the individual genera and species. When I began my study of medicine at the University of Vienna in 1922, I already had a strong interest in evolution, but I believed that paleontology offered the best avenue for pursuing that interest. But then I was fortunate to encounter Ferdinand Hochstetter (Figure 2), a teacher versed in both comparative anatomy and in comparative embryology, and I soon saw that the study of similarities and differences among living organisms would be at least as good a means of reconstructing evolutionary trees as the study of fossils. I also learned from Hochstetter that

Figure 2: *Ferdinand Hochstetter* (1861–1954).

Figure 3: *Oskar Heinroth* (1871–1945).

the individual development of a living organism—its ontogeny—yields valuable insights into the evolutionary history of the species. (In this regard, the work of Heinroth has special significance. He reared young birds from the egg whenever possible and photographed virtually all the bird species found in central Europe.) The thorough training Hochstetter gave me in comparative anatomy and embryology led to a major insight, which was to determine the whole course of my life's work: that it is possible to apply the methods of comparative anatomy and embryology directly and without modification to the study of animal behavior.

Soon afterward, I met Oskar Heinroth (Figure 3), who had reached the same conclusion many years earlier. Not only had he discovered that behavior patterns can characterize species, genera, and even orders just as reliably as anatomical characteristics (dental formulae, individual bony structures, bodily dimensions, and the like) but he had also come to the realization that ontogeny can often yield highly significant information about the evolutionary history of a particular species. And he had made all these basic discoveries through

5

Figure 4: *Charles Otis Whitman (1842–1910).*

his work with the same group of birds, members of the duck family. Only much later, long after Heinroth and I had become close friends, did we learn that the real pioneer in the field of comparative behavioral studies was Charles Otis Whitman (Figure 4), who had conducted the same kind of observations before Heinroth's and well before mine. His research had involved another group of birds, the pigeon family (Columbidae), but he had reached exactly the same conclusions.

Whitman has remained almost unknown to professional psychologists. I once attended some lectures in psychology given by my respected teacher Karl Bühler, who played host to many American psychologists, and I undertook to ask each of them in turn whether they had heard of Whitman. Not one knew of his work. Many years later, I chanced to meet his son, Charles I. Whitman, a successful businessman, and even he had no idea of the significance of his father's work. All he could tell me was that "he was crazy about pigeons."

This is an appropriate place to comment on animal lovers, to whom scientists often apply the words amateur and dilettante in a derogatory sense. Yet the word amateur comes

6

from the Latin *amare* (to love) and dilettante from the Italian *dilettarsi* (to delight in something). Nowadays, it is regarded as modern to set experimentation above observation (no matter how assumption-free) and to see quantification as a more important source of understanding than description. We tend to forget that description is the foundation of all science. I do not mean to question the value of experiments, but observation must come first, in order to generate the questions for experimentation to answer. The emphasis on blind, quantitative experimentation without prior observation is based on the erroneous assumption that scientists already know the questions to ask about the natural world. A theoretical interest and patience are not the sole qualifications for arriving at the principles that govern the patterns of social behavior in higher animals. That aim can be achieved only by those of us whose attention is riveted on the behavior of our animal subjects because of the great pleasure we, amateurs and dilettantes, take in our work.

Objectives and Methodology

The Analysis and Representation of Systems

A system is an entity made up of various interacting components that must all be present if the character of the system is to remain intact. Teaching and research face the same problem in arriving at an understanding of a system, as the following example illustrates. In attempting to explain the operation of a typical combustion engine to somebody who knows nothing about engines, we can start wherever we like. We could begin by saying, "The descending piston sucks the explosive mixture from the carburetor," although it is obvious that the hearer would not be able to make any sense out of these words. The hope is that the hearer will create a mental

7

slot for each word and can later fill in the slots with appropriate concepts. The same principle can be used to construct a flowchart in which an empty box is left for every function that is unclear. The point is that a preliminary sketch of the entire system is necessary in research as in learning, but that the sketch must leave space for certain functions within the system which will be comprehensible only when the other functions are understood. Only when one understands all the components that supply the engine's flywheel with energy can one see where the piston gets the impetus to descend. The functioning of a system can somewhat loosely be defined in terms of an understanding of all its component parts. Our definition will be imprecise, for we can never fully understand all the components of even such a simple system as the combustion engine; yet that is no reason to give up trying to analyze the entire system.

We confront just this problem in examining a system whose nature is determined by the interaction of a number of subsystems. In the second part of Goethe's *Faust*, Helena says: "But I am talking to the wind, for in vain does a word strive to construct and create form." Likewise, a linear string of words cannot provide a satisfactory representation of a system. We define a system as a varied set of structures or functions that interact with each other but that constitute a whole sufficiently demarcated from its environment as to permit recognition of a common function. It is in this sense that we can appreciate Paul Weiss's witty aphorism, "A system is anything cohesive enough to deserve a name."

In the construction of a flowchart, as in the analysis of a system, our understanding always progresses from the whole to its parts, not from a part to the whole. Before we can comprehend the functions of the individual components of the combustion engine, we must understand the function of the whole—that the engine is a source of power. The progression from the whole to its parts also applies when it is an organic entity we are seeking to understand. The art of analysis in such a case lies in the isolation of the individual components without losing sight of the larger entity and its function.

Even an approximate understanding of the interacting components in a system can produce a major advance in research, leading us closer to the stage at which it is appropriate to begin experimentation or make measurements. Ruprecht Matthaei, in *Das Gestaltproblem* (The Problem of Gestalt), draws a parallel between a researcher and a painter: "A rough sketch of the whole is progressively filled in, as the painter develops all the individual parts as much as possible together. The picture appears complete at every stage of its development—until the finished painting stands before us as a self-evident work of art." Otto Koehler has referred to this method of working as "analysis on a broad front." In biology, progress in research and teaching must proceed *from* the whole system under investigation *to* its individual components.

Organic Systems as a Subject of Research

Researchers who conduct experiments with living systems must always remember how easily the natural mechanism can be disrupted by their inquisitive tampering. Fritz Knoll gave a thorough account of this fundamental problem as early as 1926, and Eckhard Hess later gave a set of methodological guidelines in his published discussion of the "nondisruptive experiment." The vulnerability of a system to human intervention increases with the system's degree of complexity and differentiation. The higher animals are themselves extraordinarily complex systems, but the societies in which they live are still more complex, and the social life of human beings is the most complex system we know of. It is well known that many animals show, in a simpler form, parallels to human social life, parallels that may therefore be easier to understand. The vulnerability of animal societies to disruption is not an insurmountable difficulty, but neither should it be underestimated.

The most exact, but also the most demanding, method

9

for investigating the social behavior of higher animals is undoubtedly observation under natural conditions. This requires the painstaking and extensive habituation of the animals to the human observer. Jane Goodall managed to win the acceptance of a social unit of chimpanzees in the Gombe Stream Reserve, although it took almost a year before the flight distance of the chimpanzees decreased sufficiently for her to begin her observations. But the number of observations she was eventually able to make richly rewarded her investment of time and patience. Hans Kummer used similar methods of observation in studying baboons, as did Dian Fossey with mountain gorillas and Anne Rasa in her investigations of dwarf mongooses.

Another, less demanding, approach is to reintroduce hand-reared animals into their natural environment, in a setting in which they can be observed comfortably at close quarters. But this is possible only with animals in which traditions do not figure prominently in their social behavior. Katharina Heinroth's attempt to integrate hand-reared young baboons into the long-established baboon troop at the Berlin Zoological Gardens was an utter failure. The hand-reared individuals apparently behaved in a manner that clashed with the tradition of the troop, and they were repeatedly rejected.

With birds, differences in tradition between individual social groups do not play an important part because most of their behavior patterns are phylogenetically preprogrammed. It is perfectly possible to form social groups that include hand-reared individuals and still exhibit more or less normal behavior—though only after the group has led an independent existence for a considerable time. This is clearly shown in the case histories of our geese as carefully recorded over the past thirty-five years. The oft-voiced criticism that the behavior of geese may be distorted by their relationship with human beings is unjustified. To the contrary, the minor aberrations that sometimes occur are themselves valuable material for analysis.

The assumption that a colony of hand-reared greylag geese kept in a natural environment will exhibit virtually normal behavior is confirmed by the fact that whenever wild greylag geese with no human contact whatsoever have joined the colony, they have behaved exactly the same as its long-term members.

The task of investigating the internal workings of an extremely complex social system understandably must extend over a long period of time. I know of only three longitudinal investigations of the social systems of nondomesticated higher vertebrates that have had an adequate quantitative and temporal framework: Jane Goodall's study of chimpanzees in the Gombe Stream Reserve in Tanzania, the study of Japanese macaques (*Macaca fuscata*) conducted by Masao Kawai and S. Kawamura on the Japanese island of Koshima, and, finally, our own study of the greylag geese. Although my first small colony of greylag geese existed only from 1936 to 1940, some of the individual observations from that period are sufficiently important to be included in this book.

The present goose colony in Grünau was initially established on the ponds of Schloss Buldern in 1949, was moved to the newly founded institute at Seewiesen, near Starnberg, in 1955, and finally was successfully transferred to Grünau, in the Alm Valley of Upper Austria, after my retirement from the Max Planck Society. The colony now contains about 150 geese. The number varies, because some pairs breed elsewhere and spend only the winter with their offspring in the Alm Valley.

The Structure of This Book

The ideal procedure from an instructional point of view would be for me to lead the student or reader along the same path our research took in reaching the present array of results. Such a path is, however, a practical impossibility; it would be too time-consuming to repeat all the research. That is why most textbooks set off in the opposite direction; the first section is almost always general, while the second deals with specifics. By contrast, research workers tend to follow an inductive process, starting with specific observations and abstracting generalizations from them. So the usual procedure of textbooks accustoms the student to an approach that is in conflict with the spirit of inductive research, an approach that requires the formulation of a hypothesis and then a search for examples supporting it in observed reality. But the organic world is so diverse that, if one looks hard enough, one can find deceptively convincing examples for the most far-fetched theories.

The aim of this book is to render understandable an extremely complex organic system, the social behavior of an animal species, and also to give an account of that species' relationship with the ecology. To achieve this aim, I will attempt a modified inductive presentation: first giving illustrative descriptions of certain greylag geese, based on our extensive knowledge of them, and then discussing the theory of instinctive behavior. This theoretical section will encompass the behavioral repertoire of the greylag goose, including the displays and vocalizations that constitute the foundation of its social structure.

Such a strategy corresponds closely to the procedure that the philosopher of science Wilhelm Windelband has said is characteristic of all scientific investigation. The ideographic, or pictorial, first stage is confined to the description of existing phenomena. The next stage introduces systematic order to the described material. Finally comes the nomothetic stage, in which general principles are abstracted from the systematic framework. It is no different from the process the painter goes through in the description by Matthaei that I quoted a few pages earlier. Nevertheless, even if this sequence of stages is carefully followed, it is almost impossible to resist pressing ahead to aspects that the reader cannot yet know. I do this when I consciously try to work without the various interpretations that I have extracted over the past seventy-five years. My simplest descriptions of the behavior of my animals are unavoidably influenced by the systematic order from which I know general principles can be abstracted. So in describing the behavior of my geese, I will not avoid concepts that may at first be incomprehensible or only intuitively understood. I beg the reader to treat such concepts in the spirit of my discussion of the flowchart and the example of the combustion engine. In that spirit, one can read the following sections as easily as the preceding ones.

Martina

My long-term imprinting on ducks did not erase my strong interest in greylag geese. In beginning with the life story of my first goose, I am trying to reflect the development of my own knowledge of greylag geese. We know today that her life story was quite different from that of the typical greylag goose because of her rearing, which did not meet the natural requirements of the species and involved a great deal of stress. Nevertheless, this goose showed in her development many of the behavior patterns that can also be observed in a greylag goose growing up without interference and under natural conditions.

Martina's Infancy

Although the events I am about to relate took place fifty years ago, my notes are so reliable and my memory so vivid that I believe I can draw an informative picture of the life of a greylag goose by telling this story of my first goose. Incidentally, she was named not in honor of Saint Martin but after a friend. I wrote twice to Count Esterhazy requesting some eggs of the greylag goose, but when he did not answer, I turned to an illegal source. My friend Otto H. Antonius, director of the Zoological Garden at Schönbrun, had obtained live animals from this source, so I had no hesitation in following his example. In order to be absolutely sure of getting the eggs, I

Figure 5: *A gosling with its downy plumage.*

approached two suppliers, and unexpectedly both of them complied.

Finding myself with twenty greylag goose eggs, I placed ten of them under a domestic goose that was known to brood reliably and the rest under a turkey. My intention was to let the domestic goose lead all twenty goslings, which probably would have worked. By good fortune, however, it happened otherwise! After the first gosling had hatched and dried, I was unable to resist the temptation of removing the delightful creature from under the foster mother and taking a closer look (Figure 5). As I did so, it gazed at me and soon began to utter its single-syllable lost calls. Because of my previous experience with domestic ducks, I was able to interpret this correctly as crying (Plate I), and I answered with a few comforting sounds. The gosling promptly turned to face me directly, stretched its neck, and uttered a multisyllable *vee-vee-vee-vee*. It was clear to me that the transition from single-syllable piping to the multisyllable *vee* calls was a shift from distress to pleasure and that the stretching of the neck was a greeting gesture.

Who would not have wanted to see a repeat performance of this transition from desperate crying to happy greeting? I therefore remained motionless and silent until the gosling began to cry again, so that I could again comfort it with friendly sounds. Eventually, however, I had enough of this baby-sitting. I placed the gosling back under the wings of the brooding domestic goose and started to leave. I should have known better.

I had moved only a few paces when a questioning, whispering call emerged from beneath the white feathers. The domestic goose followed the standard program by replying with the contact call *gang gang gang*. But the gosling did not respond by calming down, as an infant not exposed to the special experience of my baby goose would have. Instead, it emerged from beneath the foster mother's belly in a determined fashion, tilted its head to look up at her with one eye, and ran away from her crying loudly. The single-syllable distress call, which is characteristic of many precocial birds, is extremely plaintive and almost always evokes sympathy from a human bystander. The poor gosling stood halfway between me and the domestic goose piping loudly and continuously with its neck stretched upward. When I made a slight movement, the gosling ran toward me uttering the *vee-vee-vee-vee* greeting with its neck extended forward. I was not then aware of the characteristic irreversibility of the imprinting process in geese. I therefore grasped the gosling and placed it under the belly of the white goose a second time. But it immediately ran after me again. It could not stand properly on its feet and rested on its heels, was rather unstable even when walking slowly, and waggled from side to side. At the peak of its fear, however, it was perfectly able to perform the movements necessary for rapid, propelled running. Many precocial birds, particularly gallinaceous birds, can run before they can walk or even stand still.

Understandably, I was moved by the spectacle of this poor infant running after me and crying. Although it stumbled and sometimes rolled head over heels, it followed me with a surprising speed and determination that carried an unmistakable message: it regarded me, not the white domestic goose, as its mother.

Because it takes less work to raise ten young geese than to raise one alone, I took all the goslings that hatched under the domestic goose and replaced them with the ten that hatched under the turkey. At first, though, I tried to raise Martina separately from the others, assuming that this would forge a particularly strong bond between us. I also believed that young greylag geese hatched by a domestic goose and raised in our garden might be less inclined to fly away to the nearby Danube or elsewhere.

Both assumptions proved totally wrong. It soon became obvious that the separated gosling was following me less reliably than the entire group of siblings was. On her own, Martina was always somewhat nervous and ready to flee. She tended to produce distress calls and seemed to be generally less active than the goslings in the flock. Any living organism that belongs to a highly organized social group, as the greylag goose does, needs the feedback of social stimuli to maintain the normal state of general arousal.

Martina gradually became better accustomed to following me, even in the absence of her nine siblings, and that was fortunate, because I could take only one goose with me on long trips in my collapsible dinghy. She rapidly learned to climb aboard when she got too wet during an extended period of swimming, which happened frequently at the beginning. Later, when the down feathers developed and she became properly waterproof, she traveled in the boat with me only occasionally.

Since this was long before I learned from real greylag goose parents what and how much one can expect of a following flock of goslings, I made a number of mistakes, particularly with Martina. The coarseness and cruelty of the errors did not become apparent until much later. For instance, the way from our house to the Danube led along the main street of the village of Altenberg, which was always full of inquisitive people, dogs, and noisy vehicles. Even the most obedient goose was not up to the task of following me along this path, and so I used to clasp Martina under my arm and carry her along the busiest stretch. Obviously she was not put off by this procedure and would walk up trustingly, uttering greeting calls, for me to pick her up. Nevertheless, it is now my belief that this treatment not only disrupted Martina's development but also, paradoxically, led to an early onset of sexual maturity.

My assumption that the goslings remaining with the domestic goose would be more attached to their home base than the geese I was leading around on long journeys also turned out to be completely wrong. Only a few individuals in the flock that made excursions with me on the Danube over distances of several kilometers ever strayed. By contrast, many of the goslings that had stayed at home strayed the next autumn, especially in foggy weather.

Anecdotal Observations of Martina's Behavior

I shared my bedroom with Martina for more than a year, and this close cohabitation allowed me to make numerous anecdotal observations. Several of them are worth mentioning, even though they do not fit directly into the context of this book. Oskar Heinroth had noted the remarkable achievements of anatids (members of the duck family) in the *transposition* of their knowledge of spatial relationships. They demonstrate this ability when, having learned a given pathway only from above, while flying, they must find the same way on foot, or, conversely, when they must find the correct orientation by flying along a route they have previously covered only by walking or swimming.

My ignorance of this astounding ability once caused me several hours of serious anxiety and strenuous searching for

Martina. The path home from the spot where I used to land on my return from trips on the Danube passed first across a large meadow of some 1,000 square meters and then through a dense willow wood of middling height. I had used the meadow on many occasions for practice flights, by crouching down and then leaping up and running as fast as I could into the wind while uttering flight calls. I had learned the technique of crouching from my jackdaw, Tschock, and was not yet aware that it is unnecessary with geese. While Martina's wings were still short and she was just capable of flying, I had thought I could shorten the trip home by letting her fly across this meadow. After landing from the river, I would allow her a short pause to preen herself, as I already knew that there is no way of persuading a goose to follow when the feathers need preening. This time, after the preening pause, I gave all the flight signals I knew and ran off. As planned, Martina took off and flew after me. Naturally, she quickly overtook me, but she had risen higher in the air than either of us had intended. She crossed the meadow toward where the edge of the wood loomed. Then she started to brake but saw that it was too late, that she would crash into the trees, and so she accelerated again. With a woodpeckerlike swoop, she just cleared the top of the wood—and disappeared. Our garden is situated on the western edge of the Vienna Forest, on its outermost slope. It is separated from the plain of the Tullner field by a wall some four meters in height that borders a country road leading down to the floodplain of the Danube across another steep slope. In addition, there is a row of tall pine trees along the bottom end of the garden. It seemed virtually impossible to me that my goose, barely able to fly, could cope with all the changes in height presented by this series of obstacles. (I knew that night herons, which are also open-country birds, initially find it difficult to fly upward over sloping ground; they drift sideways and reach their home base by flying painstakingly from tree to tree.) Accordingly, I did not bother to look for Martina at home, but wandered up and down the edge of the floodplain, going as far as the next village. Finally, when it was almost dark, I gave up in despair and went home. There I found Martina, standing on the mat outside the door, and she greeted me in great agitation. For a wild animal, the absence of a companion from the usual place at the proper time almost always signifies a catastrophe, so it is not anthropomorphism when I say that Martina must have been extremely worried about me by then.

The remarkable thing about the performance of this young goose was her oriented localization of a goal along a route that she had not previously followed in just this manner. She must have built up an overall picture of the locality while moving along the village street—despite the fact that she often was carried—and passing through the forest and across the meadow. She was able to recognize the overall picture from high in the sky and so find her way. Also remarkable was the way she solved a detour problem: in order to get around the obstacle of the pine trees, Martina must have flown a minimum of two to four circles, since the ability of a goose to rise higher while flying is limited and Martina, as I have noted, had not yet achieved her full flying capacity.

Several years later, an equally impressive feat of transposition—in the opposite direction—was demonstrated by my gander Viktor, who had been led by the domestic goose I mentioned earlier. One very foggy winter afternoon, I happened to notice that Viktor was missing. When a young bird is missing, there is a good chance that it has lost its way and will eventually turn up at home. But the absence of an adult gander, especially just before nightfall, is cause for great anxiety. During foggy weather, the geese sometimes land on the country road because its lighter hue stands out against the surrounding green. So I decided to look in the lane that runs past our garden and leads downhill toward the Danube at right angles to the country road. I was just in time to see the gander stride quickly along the road, turn into the lane, and hurry past me into our yard. Before then, he had walked out through the yard door and along the village street with his siblings, but he had not seen the country road and its relationship to our lane except from the air. One must know how hesitant and anxious wild geese are when crossing unknown terrain in order to appreciate the confidence with which this gander was heading toward his goal. With nightfall approaching, Viktor had obviously had difficulty landing in our garden—practically a wood—in the thick fog. He had decided to land on the wide, light country road instead. And he had known exactly where to land and how to continue on foot, rather than taking off again and risking an additional landing.

Such skill in dealing with complex spatial relationships represents a major achievement for a bird that under natural conditions inhabits open terrain and expanses of water bordered by plains, and it undoubtedly involves considerable stress. Martina displayed both these features, the skill and the accompanying stress, particularly when coping with spatial problems inside our house. At first I simply carried her upstairs, but later I allowed her to walk up the wide staircase between the first two floors and then up the narrow spiral staircase leading to the attic. That was admittedly a strenuous undertaking for the young goose, but not as hard on her nervous system as when she was picked up and carried. She had more trouble going down the stairs, but that disappeared as soon as she could fly. I then began placing her on the windowsill until she learned to fly through the window without hitting the frame—not easy, as the window was narrower than her wingspan. She would flutter up toward the ceiling and then half-fold her wings and plummet through the window without touching the sides. Since the greylag goose is, after all, an inhabitant of open country, this aerobatic feat never ceased to impress me.

Animals are creatures of habit. Because their capacity for abstraction is far less developed than that of humans and because they lack the ability for causal reasoning, their self-conditioning must serve both these purposes. The rigidity of path conditioning in greylag geese is illustrated by one observation I made with Martina. When she was still a small gosling, I managed to induce her to come through the front door, across the hallway and antechamber, and into the large central room. Alarmed by these new surroundings, she ran to the large window opposite the entrance. (Frightened birds always move toward the light.) But the convex base of the staircase that she had to climb when following me up to my room was close to the entrance to the room and at least three-quarters of the room's length away from the window. I had to entice the gosling away from the window and back toward the staircase, where she finally climbed up onto the first step on the left side. On entering the room the next day, she ran toward the window again, but immediately consented to turn around and climb onto the first step of the staircase. For many days afterward, she continued to make this sharply angled detour to the window, although the angle of the detour gradually flattened out until she was making a simple right-angled turn onto the first step.

One evening at about this time I forgot to bring Martina inside. When I finally remembered to open the door, she was standing on the mat in a somewhat agitated condition, and she rushed between my legs into the house. She climbed onto the first step on the right side and ran up the stairs by the most direct pathway. When she reached the fifth step, however, she did something quite peculiar. She froze in a posture of extreme alarm and uttered the alarm call. Then she turned around, came back down the five steps, and hurriedly followed her detour to the window, like somebody performing a bothersome formality. Having done this, she again climbed the staircase to the fifth step, stopped, and relaxed. She shook herself, performed a greeting, and calmly continued up the stairs. For a living organism that totally lacks any capacity for abstraction or causal reasoning, it is undoubtedly a good behavioral strategy to cling slavishly to a procedure that has proved to be safe and successful.

I had accustomed Martina from the outset to follow me when no other geese were present. I had originally hoped to train other members of my flock to follow me alone, but that proved impossible. The cohesion between siblings is so strong that they show great agitation when just a few are missing and tend to utter distress calls, display alarm postures, and flee. Individuals are not at all willing to follow a human leader. I therefore began to lead the geese down to the Danube as a group, although I had to circle around the village to avoid its many frightening features, which took a great deal of extra time. On the other hand, it was not hard to train the geese to swim after my kayak. I noticed that when swimming they kept far closer to the bow of the boat than they did to my heels when walking, and I realized that young greylag geese measure their distance from a substitute parent by the angle formed between the top of the person's outline and the horizon.

Their recognition of people turned out to have nothing to do with clothing. If I had been naked and suddenly appeared fully clothed, it did not affect Martina in the slightest. On the other hand, she was momentarily afraid of me if I went into the water so that she could see nothing but my head. She would display discomfiture, turning toward me and away again. Then she would have a sort of flash of recognition and would greet my face by swimming right up to me while uttering intensive *vee* calls with her neck outstretched. It was more difficult for her when my wife climbed into the collapsible dinghy and I swam alongside. At first Martina was quite relaxed and took up her usual position near the bow, close to the left paddle. But when she looked up and saw the upper part of my wife's body instead of mine, she became thoroughly frightened, diving down and resurfacing some distance from the boat. After frequent switching of the occupant of the boat, however, she eventually overcame her fears.

Another interesting observation with Martina came in the context of the problem of recognizing individuals. Following a lengthy day trip on the Danube, I had alighted from the boat at our usual landing place. I was just about to get dressed, and Martina stood close to me on the bank preening herself. Suddenly she stretched her neck and uttered a distance call. One quickly learns to recognize the direction in which a bird is looking from the position of its head and eyes. When I followed Martina's gaze, I saw drifting along the opposite bank of the Danube a white kayak in which sat a man with a beard who closely resembled me from that distance. I intuitively understood that Martina had taken this man to be me and that this misunderstanding was in no way affected by the fact that I was standing just a few paces from her. Although I tried to attract her attention by moving and making calls, she took off and flew straight across the river toward the other boat. She was only a few meters from the stranger and preparing to land when she realized her mistake, uttered an alarm call in great fright, and flew sharply upward. Martina did not fly back to me after this incident, but instead flew straight back to the garden of our home.

Like Martina, my other young geese were extremely reluctant to land in a strange place. I saw this behavior repeatedly later on, in Buldern, Seewiesen, and Grünau. During excursions on which the geese were taken farther than usual from their home port, their response to any scare big enough to make them take off was to abandon their human leader and return home. Since I often took Martina on excursions covering several kilometers, any such fright always led to an unwanted termination of the outing. In most cases, however, she would not return home immediately but would fly as far as our garden and then back to me, circling several times and showing a clear intention to land, only to return and land in the garden at Altenberg after all. Because she climbed quite high in the sky when flying so far, I could see from a great distance away that she really did fly right to our garden. It was my impression that she abandoned the possibility of returning to me only when she had managed to localize reliably the landing place at home.

24

In the early spring of the following year, Martina paired with a gander from the flock led by the white domestic goose. This was unusually early for a greylag goose to pair. Normally, conclusive pairing does not occur until the goose's second spring. I know from later observations, however, that the processes involved in the formation of this pair were completely normal, and because I have a particularly clear recollection of the details, due to my close relationship with Martina, I shall describe them here.

The first thing I noticed was the frigate posture shown by the gander, in which he swims high in the water, wings raised slightly, the rear of the body extended, and the neck held in an elegant arch posture. The overall body posture is somewhat reminiscent of the display posture of the mute swan, but I do not know whether the two postures are homologous. The gander turns his flank toward the goose he is courting, and may even turn full circle if he is swimming on the spot and the goose passes by. All this came to my notice because Martina was still accompanying me on many occasions, which meant that the courtship display of the gander was oriented in my direction as well. Less abruptly, and therefore not immediately noticed by me, they began to perform the strange pattern of parallel walking, in which the gander accompanies the goose step for step, literally copying her every movement. If she stops suddenly, he freezes with one foot in the air. Further, he will follow her to places that are known to fill him with fear. A gander in this state often gives the impression of not being "all there." He will unselectively attack anything that crosses his path—not only other geese or various creatures that he is normally afraid of, but also such pseudo opponents as a watering can standing in his way. Martin, as we decided to name him, showed no fear of our testy old peacock or of me. While in this exalted state, he even followed Martina through the door of the house and up the stairs—an achievement that is virtually unheard-of with greylag geese. His extreme excitation was obvious from the pronounced trembling of his neck and from his bulging eyes. I can still visualize him standing in our attic room, his feathers excessively flattened and his neck appearing quite thin, trembling in fear and repeatedly hissing loudly. Suddenly a door banged shut in a room nearby, and that was too much even for a courting gander. Martin took off blindly and flew into a glass chandelier. The chandelier shed some of its ornaments, and the gander lost one of his primary feathers.

Sadly, Martina and Martin soon disappeared. It may be that they were unable to find a suitable nesting site in our overcrowded garden, although I now think it more likely that they flew off simply to escape from the constant stress.

On Normality

Almost two hundred years ago, Johann Wolfgang von Goethe firmly believed that an "archaic plant," containing all the organs found in modern plants, existed and would eventually be discovered somewhere. The great poet forgot that a type is only an abstraction, an indispensable aid to reasoning but not existing in reality. *The* plant, which Goethe hoped to find, does not exist and never has. What does exist is a great variety of individual plants from whose structures it is possible to abstract a general type.

Similarly, *the* greylag goose, discussed by many authors (including myself), does not exist. There is a multitude of real geese that can be identified unequivocally as greylag geese. They all share a number of characteristics that are genetically determined, characteristics that include a set of programmed motor activities, the so-called instinctive motor patterns. These are laid down in the genotype and represent the complete repertoire of genetically determined behavioral capacities, which is the subject of our research.

The concept of the genotype is as indispensable for our studies as the concept of the ideal healthy human being is for the physician. The ethologist can approach the behavioral genotype solely through the study of how certain noninherited variations and the interactions between instinctive motor patterns influence the behavior of individuals. In other words, we can investigate the genotype only through the phenotype, or the observable expression of the genotype, which is influenced by all these factors.

An inventory, or repertoire, of the behavioral systems

programmed into each individual is called an *ethogram*. Yet every behavioral system can be modified to some extent by external conditions, and each modification can have a significant effect on other systems. Although every pattern of instinctive behavior has its own spontaneous activity and "a seat and a vote in the great parliament of the instincts," the role it plays in the life of the individual is not always fixed. We know of cases in which certain instinctive behavior patterns have never been performed in the life of a individual. We know of other cases in which certain patterns have developed much earlier than normal. A case that combines in perfect balance all the characteristics of inherited behavior can indeed be formulated, but it will never be found expressed in any actual individual.

Heinroth's starkly abstracted picture of the life history of greylag geese goes roughly as follows: After the end of the first year of life, the goose and the gander break away from their parental family. In the course of the next year, usually in the early spring, they fall in love, perform the triumph ceremony, and subsequently live as an indivisible pair. If we examine our carefully compiled behavioral records in search of such an ideal case, however, we find not a single one. Every individual biography has its flaws. Many years ago, I asked Helga Mamblona-Fischer to sift through our hundreds of records and pick out any in which pair formation and pair stability approximated the type defined by Heinroth. When I clearly showed my disappointment over the tiny sample that resulted, this provoked from Helga the classic comment: "Geese are only people, after all."

There is a reason that the records showing near-normal pairs of greylag geese all date from the more recent history of our colony. In the early years, the behavior of our geese suffered much more disruption than it did later on, particularly because of the limited choice of mates afforded by the small colony. The fact that I was able to observe an approximately normal process of pair formation with Martina and

Martin was due to the lucky coincidence that they had been separately led—by me and by the domestic goose. They did not recognize each other as siblings, and therefore the incest taboo did not operate.

There is a great number of different sequences in the behavioral systems that are common to all greylag geese. Accordingly, it has been no easy task to select from our extensive records those individual goose biographies that best illustrate the innate behavioral structures of the greylag goose, or, more correctly, the greylag pair. But the fact that the repertoire of the greylag goose's behavior is known with great precision is a comfort to researchers. No new patterns are invented, as is the case with advanced mammals, most notably the primates. There is, so to speak, a given framework within which the behavior can develop.

The Habitat and Life Cycle
of Our Geese

The Move from Seewiesen to Grünau

In 1973, when I retired from my post as director of the Max Planck Institute for Behavioral Physiology, which is located on Lake Ess, near Starnberg, in Upper Bavaria, I was planning to carry out a dynamic longitudinal study of greylag geese, and for that purpose we had established a resident flock of the birds. Our method, a long-term investigation of the social structure of tame birds living in complete freedom, had already proved rewarding and showed great promise. It was clear that the value of such a study would increase dramatically with its duration.

I was therefore extremely grateful when the Max Planck Society generously provided funds not only for the transfer of the colony of greylag geese to Austria but also for the initial costs of maintaining the colony and continuing the research. In Austria, we were warmly welcomed by the Ministry for Science and Research and by the Austrian Academy of Sciences.

A research center was found for us, at nominal rent, by the House of Cumberland—the Auingerhof, a beautiful old two-story millhouse, completely furnished (Plate II). Under the presidency of Chief Forestry Officer Karl Hüthmayr, the Cumberland Foundation constructed a pond environment

(Oberganslbach in Plate II) upstream from the local wildlife park, along with three small heated wooden huts for the caretakers of the hand-reared geese.

In order to strengthen their bonding to human beings in preparation for the transfer, the eighteen geese we reared in 1971 were tended until late fall by their foster mother, Sybille Kalas-Schäfer. In 1972, three small flocks were hand-reared, with the aim of investigating rank order among siblings. With these geese as well, we emphasized bonding with humans in order to facilitate the transfer. In the spring of 1973, the geese reared the previous two years were still clearly attached to humans. Along with four flocks that were later reared to fledging in the Alm Valley, these geese formed the nucleus of the resident colony that we established in our new home. We also took as many other hand-reared geese as we could, especially parents that were leading offspring.

We chose the early summer of 1973 as the best time to transfer the geese, because their primary feathers were molting and they were unable to fly. We were thus able to avoid the further stress of cropping the primary feathers. The hand-reared geese that were transferred after they had started to fly had no problem attaching themselves to their former foster parents or, indeed, to any other humans in their new environment.

At the time of the move, the huts at Oberganslbach were not yet completed, so the three coworkers who were leading flocks of geese had to camp out with their charges in one of the feeding sheds in the wildlife park. This area, which bordered on the southern end of the grounds at Oberganslbach, became something of a resettlement center for the goose colony in the Alm Valley. The adult geese were initially housed in an aviary near the Auingerhof, but we soon moved them to an aviary in the wildlife park because it provided vegetation as well as shade. When we released the geese, they showed no close attachment to this site, and the fact that at least some of them eventually settled in the facility at Grünau is almost entirely due to the efforts of Sybille Kalas-Schäfer, who looked after them there.

A number of problems faced the geese in their transfer from the biotope of Seewiesen, partly due to the different grazing conditions and partly due to the unfamiliar terrain. It took a while for the geese to become accustomed to the tougher grasses and herbaceous plants in the mountain valley. For the first few days, the young hand-reared geese ran across the meadow uttering distress calls and conspicuously searching, but not eating anything.

In Seewiesen, the geese had been used to still, clear water and an open, slightly rolling terrain. The Alm Valley, with its rushing mountain river and steep, stony banks, at first was harder for the geese to move around in. Nevertheless, they mastered the strong current of the river quite easily, although they were slower to accept the surrounding floodplain forests as choice places for cover and grazing.

By the beginning of July, all the coworkers involved in leading the geese had moved into the completed huts at Oberganslbach, and this site gradually became the center of our goose colony. However, the Kasbach pond in the wildlife park, the first large stretch of water the geese encountered, continued to be their sleeping area until they discovered Lake Alm. Our excursions to familiarize the geese with the Alm Valley were gradually extended upstream, and on July 12, 1973, we reached Lake Alm for the first time on foot (Plate II). The geese immediately felt relaxed and at home on this large expanse of water. We soon noticed that many of them— they had by now completed their molt and were again able to fly—were flying independently to Lake Alm to spend the night. This meant that we had achieved one of the most important goals of our resettlement program.

In the autumn, the human caretakers left the small huts and moved into the Auingerhof. From this time on, the geese followed their foster parents during the day on the newly completed pond system at Oberganslbach. In the afternoon they spontaneously flew to the Kasbach pond in the middle of the wildlife park. In the evening we called them to the gravel banks of the Alm, alongside the Auingerhof, and fed them.

Daily and Seasonal Changes in Site

Over the course of the year, all the geese became accustomed to spending the night on Lake Alm. During the day, however, they spent their time with humans, at Oberganslbach in the spring and summer and on the banks of the River Alm in the winter months. These habits have remained with them to the present day. A large expanse of water gives them a feeling of security, familiar humans provide them with food, and in winter the relatively dry sandy banks of the constantly flowing Alm are a welcome source of warmth.

When the time comes to find a nesting site, the flock disperses. But many of the parents return from their widely scattered nests to rear their goslings at Oberganslbach. During the rearing period, virtually all the parents that are leading offspring find their way to Oberganslbach, while the non-breeding birds move to Lake Alm by the end of May to molt. The breeding birds must spend their period of flightlessness at Oberganslbach with their goslings.

The molting of the wing feathers, during which the goose is unable to fly, varies in timing from individual to individual. Nonbreeding birds generally molt earlier in the spring, while breeding birds molt later, according to the time at which their offspring will be learning to fly. The molting of the parents is synchronized with the growth of the offspring, so that all will become able to fly at around the same time.

As autumn approaches, the geese increasingly disperse throughout the wildlife park and also make frequent visits to the meadows at the Auingerhof, where they concentrate in large numbers in the winter. That may be because the members of our team feed the geese right next to the Auingerhof during the winter. The geese arrive from Lake Alm in large bands, often forming a tightly knit flock, and land near the Auingerhof. The elevation of Lake Alm is some 100 meters higher than that of the institute, and the geese tend to maintain a constant height during their inward flight in the morning. It is an impressive sight to see the entire flock noisily arriving in tight formation.

The Daily Pattern of Activity

Greylag geese, unlike most other birds, do not distinguish sharply between periods of daily activity and nighttime rest. Only young goslings remain fast asleep in the dark, and their mothers appear to make allowance for this while sheltering them. But the adult geese in a flock are never so sound asleep that they can be taken completely by surprise at night. Also, particularly during nights of bright moonlight, one can often hear contact calls that indicate fairly intensive social inter-actions.

Geese can sleep while drifting on the water, while standing, or while lying down. The head is turned backward and the beak tucked under the shoulder feathers. When a goose sleeps standing on one leg, the beak may be pointed backward on either side of the body. Young geese show this sleeping posture even before the shoulder coverts have developed enough to provide a firm anchor for the beak. Geese can also sleep with the head held straight forward and the chin tucked in. This is regularly seen with goslings that are still in their down plumage but is less common with adult geese.

After daybreak, the geese set off to graze at promising sites, mainly grassy meadows. Whereas domestic geese simply rush to the food trough as soon as their shed is opened in

34

the morning, wild geese often fly a distance of several kilometers from their sleeping site to reach their feeding grounds.

The most important type of feeding behavior, grazing, is confined to the daylight hours, since the geese must recognize plants and collect them with an oriented grasping movement. Only one form of feeding, upending, can theoretically take place without visual monitoring. The geese show this behavior in quite opaque surroundings, but it is not known whether they do so in the dark.

The geese always bathe at the same time of day, around midday. In bathing, they perform a series of motor patterns that sometimes correspond in intensity to the most vigorous locomotor patterns but that can equally well take the form of clear-cut escape responses. It is therefore unclear whether the entire bathing process should be interpreted as a set of empty activities or as play. We have come to use the term play diving for it.

After performing their bathing activities, the geese preen themselves extensively and then they rest. Their eyes are not actually closed while they doze, but they certainly are not ready to engage in any activity. Someone leading a flock of hand-reared goslings quickly learns that they simply cannot be budged for a while after they have bathed and preened themselves.

The geese become increasingly active as the afternoon progresses, often flying short distances to and fro. It is my impression that social interactions are more frequent now than in the morning, when foraging activity takes priority.

Geese show a clear daily pattern in their reactions to possible danger. Their inclination to flee increases as daylight fades, and they try to seek out the most secure nighttime resting sites, typically alongside lakes or ponds. They often stand on the shore with their beaks turned toward the open water.

When it is nearly dark, their intentions to move on to the sleeping site become increasingly obvious. For his doctoral thesis, Alain Schmitt studied the synchronization mechanisms involved in the evening departure, and he summarized them as follows:

"Preparations for the departure flight can last up to half an hour. They ensure that all the geese in a flock are synchronized and will fly to the sleeping site in a coordinated fashion. Approximate synchronization of the flock is brought about by the decrease in light intensity, and there is a very high correlation ($r = 0.90$ for 110 departures) in flights to the sleeping site at nightfall."

Fine-tuning of the departure flight is accomplished by social factors, particularly departure calls and head movements that signal the underlying motivational state. As the time for takeoff approaches, the departure calls become louder, more abrupt, increasingly multisyllabled, and more frequent. Lateral head shaking also becomes more frequent and more intense. Finally, the head is tilted around its longitudinal axis. When combined, the shaking and tilting of the head produce a figure-eight movement. The head tilting alone is a strongly ritualized signal that becomes increasingly frequent as the motivation for departure grows. With the waning of the light, the geese gather together (clumping) and social interactions gradually become rarer. But the geese also become more inclined to move around rapidly, offsetting the clumping tendency.

With the approach of the departure flight, it becomes more and more dangerous for any goose to miss the connection, and various displacement activities thus become increasingly frequent. The geese show bathing activities, nibbling of their feet, preening behavior, and even movements indicating nest-building intentions. Then we see a new behavior pattern, straightening (a menotaxis). The geese all proceed to orient themselves with their bodies parallel and lined up with other family members, without whom they will not take off. Straightening is released by intensive departure calling or warning calls (loud alarm calls), by rapid directed movement, and by the flying to and fro of other geese. It also has an important social function that is apparent in massed flights: if the geese are not parallel when they take off, there will be collisions in the air and some individuals may be left behind.

The larger social units are commonly the first to take off. They receive more social feedback during the preparation phase than, for example, solitary individuals do.

Every evening of the autumn and winter months I watch enthralled as our large flock of geese slowly builds up its motivation for the departure flight in the dusk and then takes off.

The Life Pattern

At the end of April or the beginning of May, the greylag goslings hatch under the belly feathers of the mother. The clutch usually contains four to six eggs (the overall range is from one to nine), and the incubation period is almost exactly twenty-eight days. Shortly after hatching, a gosling will peer out from beneath its mother's belly feathers and visually scan its immediate surroundings. While still in the egg it has communicated with its siblings and with its mother, who responds with contact calls and a special hoarse call. As long as the family is undisturbed, it will remain for twelve to twenty-four hours in the nest after hatching. This respite gives any late-hatching goslings the time to complete the hatching process, and it also allows the goslings under the mother's belly to slough off the thin horny sheaths that coat all their feathers.

The gander, who has stayed a short distance away from the nest throughout the incubation period, comes closer when hatching begins and remains directly alongside the nest. Before the family leaves the incubation site—to which it never returns—individual goslings emerge from beneath the mother and investigate the nest area with somewhat unsteady steps. This is the time when the main process of filial imprinting on a leading individual takes place, accompanied by an exchange of vocalizations between the mother and her offspring. A gosling does not learn to recognize its parents individually until later, after which it confines its *vee* calls and following responses entirely to them. The goslings appear to recognize their parents by their vocalizations somewhat earlier than by their facial characteristics.

If two nests happen to be located quite close together and the goslings of both families hatch at about the same time, a fusion of the two flocks of goslings can take place. The two pairs of parents, who have no means of distinguishing their own offspring from the strangers, will sometimes fight violently. The losers will trail along behind the winning pair for a while, though never longer than a day, and then will give up. Goslings recognize their parents individually only when they have followed them for at least a day, and the parents take a little longer to recognize their offspring.

The mother goose eventually gives the signal to leave the nest by uttering louder contact calls and departure calls. In abandoning the nest, the family embarks on a demanding and dangerous journey to its summer grazing area, sometimes traveling several kilometers. In their first hours of life, cohesion among the goslings is stronger than their drive to follow the parents. Accordingly, at first one sees a clump of goslings, and only later a chain of offspring swimming or walking in tight formation behind the parents (Plate III). The offspring have a remarkable capacity for locomotion, and this serves them well, for the incubation site is not the same place as the rearing site. The securest place for nesting is not necessarily the best place to rear goslings.

Most of our goose families move downstream from the incubation area on Lake Alm to Oberganslbach, a distance of about seven kilometers. Including occasional short rest periods, the journey takes them five or six hours. On the way, the parents must keep a constant watch for predators, either from the air (ravens, hawks, golden eagles) or from the thicket (foxes, prowling dogs). They must also guide the goslings around dangerous obstacles, like the eddies and dam pools in the River Alm. It is worth noting that geese that have never followed this trail on foot but have seen it only from the air are able to find their way purposefully to the rearing area.

Not long after the goslings have learned to greet each other with slightly divergent neck stretching, they begin to take part in any conflicts between their parents and other geese. If, after an attack on other geese, the gander returns to his family uttering triumph calls, the young goslings will join in, showing exactly the same posture as the adult. When the offspring's voice breaks, shortly before fledging, the fine *vee-vee-vee-vee* that it utters during what we call family palaver is transformed into a juvenile form of the later pressed cackling. Sybille Kalas-Schäfer has dubbed this vocalization of young geese, which is directed exclusively toward the parents, fluent cackling.

When they are a few days old, the goslings engage in fighting to determine the hierarchy within the family, which then remains stable for a long period of time. As the offspring grow bigger, they begin to walk alongside or in front of the parents, rather than following, and often move off in a different direction, thus determining the route to be taken. Sometimes one or more goslings become separated from the rest of the family by a distance of several meters. If this gets them into difficulties, such as being attacked or bitten by other geese, they utter the so-called lamentation call. But the parents' readiness to defend their offspring gradually decreases as the young develop the capacity to fly.

In early summer, just before the onset of the molt, the geese show a disinclination to fly. They become extremely shy and cautious—even geese that are quite tame—shortly before the wing feathers are ready to be shed. The accompanying lowering of the threshold for stimuli that trigger escape responses is innately programmed, for birds that have been pinioned become just as shy as birds that are able to fly. Nevertheless, a number of the more human-oriented individuals in our population do not find it necessary to take refuge on Lake Alm, but go through the molt at Oberganslbach, along with the pairs that are leading goslings. But most of our geese that have not bred or have lost their offspring retreat to Lake Alm. They are difficult to spot there, for they hide in the rushes and bushes near the shore if a boat approaches.

The first feathers lost in the molt are the coverts over the wing feathers. Healthy greylag geese often lose all their wing feathers simultaneously, although it is not a sign of poor condition if they shed all the primaries first and the secondaries somewhat later. In four to six weeks, when they can fly again, the geese return from the molting area to their summer grazing grounds and form a flock with the families that have remained there.

At the age of about six weeks, when the wing feathers begin to sprout, the young geese make their first preparations for flight by flapping their wings and running forward. Soon one can see that their wings are briefly holding them aloft. Meanwhile the parents are completing the molt of their main feathers. Both the parents and the young geese can fly before the wing feathers have reached their full length. This makes the parents' flying somewhat unsteady at first—an advantage in that they are unable to perform the more difficult flying maneuvers in front of their offspring. Without careful guidance—from which hand-reared geese of course cannot benefit—accidents in flight are at first fairly common. The cohesion of the family seems to increase when the young geese fledge, at the age of approximately ten weeks. Hand-reared geese also seem suddenly to become more closely attached to their human foster parents immediately after fledging.

When the young geese are fully able to fly, the families

begin to range over a much larger area. Our geese often fly far across the plain from the Alm Valley, although they never spend the night away from Lake Alm. Some families that breed on Lake Chiem in Bavaria regularly return to Grünau at this early stage in their offspring's development. Although we have never seen a proper autumn migration with our geese, it appears that geese that breed at the northern limit of their geographical range migrate southward as soon as their offspring are fully able to fly.

In the autumn, several flocks gather together in a certain spot—usually a large lake bordered by grassy plains—and form a massive migratory flock, often containing thousands of individuals. They migrate southward, stopping on the way at a series of resting places. The knowledge of the overwintering site and its geographical location is transmitted from the parent geese to their offspring in the form of a tradition. In other words, the migratory pathway is not innately determined.

In all situations, greylag geese associate with other greylag geese. Families and larger groupings are bonded together in a hierarchically organized stable relationship characterized by reciprocal individual recognition. During migration an anonymous herd instinct keeps the large flocks of unfamiliar greylag geese together. Those that have lost their families will not fly alone, but loosely attach themselves to migratory flocks.

Innate migratory arousal becomes apparent in the winter even among our geese in the Alm Valley, all of them reared either by humans or by hand-reared parents. They do not migrate, however, because it is their tradition to spend the winter in the Alm Valley. We provide the flock with the food it needs during the cold, snowy weather.

As spring approaches, but before the breeding season begins, the bonds between parents and their offspring progressively weaken. One of Norbert Bischof's students, Helge Böttger, observed that the parents and the offspring settle down to sleep at a growing distance apart, while the parents get closer to each other. The final dissolution of the family can occur in various ways. Young males are particularly prone to leave their families, suddenly going off alone or beginning to court unfamiliar females. Sometimes the parents play an active part by rejecting their offspring. The young geese so deprived of parental protection commonly join in forming so-called sibling bands.

Greylag geese become reproductively active in their third year. One-year-olds can exhibit sexual behavior patterns, which might seem to herald the onset of sexual maturity in the second year. But we have never seen a case in which a goose of this age has actually proceeded to breed. Sometimes a one-year-old goose will pair up with its future breeding partner, often another individual in the same age group. Other young geese remain alone for a time or become members of sibling bands, not pairing off until the following spring, when they begin to breed.

The bond between the partners in a pair that successfully breeds and raises offspring can remain stable for years, perhaps for life. Females, however, are particularly vulnerable during the breeding season. Out of thirty-six clearly documented losses among our female geese, twenty-two occurred during the breeding season. Most of the losses can be attributed to natural predators, although a number of cases involved a fatal illness. It seems likely that females are more susceptible to infection because of the stress of breeding.

Due to the high loss of females during the incubation and rearing periods, roughly 50 percent of all the paired males in our flock become widowers at least once during their lifetimes. By contrast, only about a fifth of all females become widows. The subsequent history of the bereaved birds is particularly interesting. Of thirty-two ganders that lost their partners, more than half went on to form bonds with new females. Somewhat less than a third initially remained alone, and the rest formed pairs with partners of the same sex. On the other hand, 75 percent of the female geese that lost their partners formed a new pair bond, only a quarter remaining without a substitute mate.

In addition, the geese sometimes change partners. Out of sixty-one females, as many as fifteen abandoned their first mates and formed new pair bonds with other ganders. In nine of these cases, this happened after a failed attempt at breeding, and in four other cases the pairs did not breed at all. In only two cases did a change of partner occur despite the successful rearing of offspring.

The longest lifespans yet recorded in our colony were attained by a twenty-year-old female and a twenty-one-year-old male (still living). We have never observed any signs of old age, let alone of senility, among our geese.

Life Histories

Great differences in life history can exist between the partners in even the most normal pairs of geese, particularly with respect to their backgrounds. We will illustrate this with the records of particular geese.

Mercedes and Florian

Mercedes's Background

Mercedes hatched at the beginning of May 1979 and was raised, along with three siblings, by her biological parents, the pair we called the "wild ones." Her first year passed uneventfully. Then, like any normal goose, she separated from her parents at the age of one year. Following the dissolution of the family, the siblings at first maintained close contact with each other, but after the molt, at the beginning of July 1980, Mercedes returned to the flock alone. Three months later, in September, Mercedes was seen with the gander Nilson for the first time. Around the end of October, however, she tried to attach herself to a pair that was not yet firmly bonded. For approximately four months, she accompanied this pair and tried to join in their cackling, but she was never fully accepted and failed to attract the gander's attention.

In March of 1981, Nilson began to court Mercedes by approaching her with the bent-neck posture, but she did not respond to his contact calling and remained unmated. By now, she was no longer being seen with the pair she had tried to attach herself to. In the summer she was courted by a three-year-old gander. Because his courtship was not very intense, however, Mercedes did not respond to him, and no bond was formed. Then, perhaps stimulated by the behavior of this rival gander, Nilson again showed an interest in her. Mercedes did not seem to reject his approaches this time, and the two of them stayed together until the spring of 1982.

Florian's Background

Florian hatched in 1973 and was hand-reared by Brigitte Dittami-Kirchmayer, along with his five brothers and two sisters. The cohesion between the brothers seemed to be unusually strong and was noticeable even after one of them paired off. In June 1973, Florian, still an unfledged gosling, was transferred to Grünau along with his siblings and his foster mother. In 1974, while flying with one of his brothers, he lost his way and ended up in Lower Bavaria. The brother was trapped and transported to Seewiesen, but before Florian could be sent back, he managed to find his own way to Grünau and the Alm Valley.

In 1975, Florian paired with the goose Nat, but the clutch was unsuccessful and the nest was abandoned. In 1976, while on the nest, Nat was eaten by a fox. Traces in the mud on the shore of Lake Alm showed that Florian had ferociously defended her. Later in the year, Florian first accompanied the "Little Goose" and then an unringed female at Oberganslbach, and during the molting period on Lake Alm he joined up with the ganders Nikita and Gurnemanz. Then, from 1977 to 1978, he lived in a fairly tight gander band with his brothers Xaver and Markus. Various fights occurred with Xaver, who was accompanying the female Selmaweiss. In both years, Florian associated with the nonbreeding geese during the breeding season. In 1979, the year in which his future mate,

Mercedes, hatched, he was still maintaining his gander association with Markus and Xaver, but after Xaver's death (he was run over by a car in April) the association became weaker, and finally it dissolved. In July 1980, Florian returned from the molt with the mate of his brother, Valentine, but the latter won her back in a fight. Florian spent the rest of the summer alone, often isolating himself from the flock and seeking out contacts with human friends. At the end of September, he formed an association with the gander Pepino, but it was unstable—at least on Florian's part. During the next year's breeding season, he attempted to steal the female Sinda away from Blasius, a gander of the same age. On March 14, Florian was seen with Sinda near her nest, and Blasius was nowhere to be found.

At this point the diary lets us down, but apparently there was a fight between the two ganders, with Blasius emerging the victor. In any event, a very disheveled Florian arrived alone at the institute ten days later, while Blasius remained close to Sinda, shielding her from any contact with others. Florian again began to accompany his friend Pepino. The latter, however, had links with other ganders, and that led to a crisis in the friendship, which resulted in Florian's becoming virtually a whipping boy. His interest in Pepino waned. Florian was first seen cackling with Mercedes on December 27, 1981, although she was still paired with Nilson. From then on, the various relationships became very shaky, particularly after another gander began to court Pepino at the beginning of February 1982. As can be seen from the following records, a switch in pair bonding was taking place.

History of the Relationship Between Mercedes and Florian

Extracts from the diary:

Feb. 8, 1982   Pepino is courted by Serge. Nilson follows Florian in the air again. [This first aerial pursuit was unfortunately not recorded.]

Feb. 10, 1982   Serge approaches Pepino with the bent-neck posture and is very insistent [that is, he follows him step for step]. Pepino occasionally accompanies Florian, and the latter "has no idea what is going on."

Feb. 16, 1982   Violent but inconclusive fights between Nilson and Florian over Mercedes. Sinda-Hellblau [a two-year-old gander] approaches Mercedes with the bent-neck posture!

Feb. 17, 1982   Further violent fights between Nilson and Florian. Aerial pursuits occur. There is a long, violent duel with the wing shoulders, which continues until Nilson is exhausted and concedes in a lamentable state. Florian runs to Mercedes while triumph-calling (as loudly as he can after his strenuous exertions), and she responds to him.

Florian and Mercedes remained together from then on. They returned from the molt together in the summer, performed the triumph ceremony, and were regarded as firmly pair-bonded. Nilson remained alone for a considerable time, but it later emerged that he had not forgotten Mercedes. Almost exactly one year later, on January 26, 1983, the entire flock experienced an outbreak of courtship motivation, as the result of an abrupt rise in temperature, and Nilson unexpectedly began to court Mercedes again. He ran toward her from some distance away, rolling as he went, and she responded by cackling excitedly with Florian. Nilson kept close to the pair, despite the fact that Mercedes made no response to him, and the situation grew very tense. After a few hours, Nilson ran threateningly toward Florian. The latter gave a threat display in return, but Nilson attacked so vigorously that Florian flew off. Nilson took up the pursuit, and the two ganders engaged in a drawn-out aerial battle, with Nilson repeatedly trying to strike Florian with his wing elbows. Meanwhile, Mercedes was running to and fro seeking her mate and uttering distance calls. Finally Florian landed nearby, completely out of breath. When he had recovered somewhat, he moved toward Mercedes. She approached him, and they began to cackle quietly together. Nilson stood motionless nearby. Although Mercedes showed no interest in Nilson, Florian shielded her against him, but Nilson did not attack again.

Nilson flew in with the flock the following morning, but Florian and Mercedes did not show up until late in the morning. We had been unable to observe whether there had been further conflict during the night. In any event, Nilson made no further attempt to win Mercedes back. (He found himself a young female the same year and paired with her the next spring, but he disappeared during the breeding season. He presumably fell prey to a fox.)

Although it was Nilson who had won the fight, Mercedes remained firmly attached to Florian, an indication that the bond between them was particularly strong. The two have been a very good pair to this day. Indeed, they hold a record in hatching six goslings from six eggs on Lake Alm in 1983, leading them without loss over a distance of seven kilometers to our ponds at Oberganslbach, and successfully rearing them all to fledging. In 1984, they bred at the same nest site and raised five goslings to fledging from five eggs. In the late winter of 1984–85, this family of seven geese could be recognized from some distance away because of its close cohesion, and it occupied first place in the rank order of our goose colony. In 1985, we reduced the clutch incubated by Mercedes to four eggs. Three healthy goslings hatched from these eggs, with the fourth found dead in the nest. At the beginning of 1986, the parents and these three offspring were still a closely knit family. Because the parents did not breed that year, one of their offspring from 1985 rejoined them after the molt. In 1987, they again produced no offspring.

At the beginning of every year in which they were still a family with their offspring from the previous year, Mercedes and Florian as a pair occupied a very high position in the rank order. They would decline somewhat in rank later in the year, especially if they did not breed. But in the winter the pair would again achieve a high rank, with the support of their grown-up offspring. Florian seems to have little inclination toward belligerence and is not particularly aggressive, but he sometimes exhibits an extreme dislike of certain individuals and repeatedly engages in aerial battles with them. Our impression is that some degree of rivalry with particular ganders has carried over from an earlier period, since the other ganders, including his own brothers, are almost all from the same hatching year as Florian. He is quite often the loser in the fights that continue to flare up now and again. The dominance-submission relationships are difficult to assess overall because Mercedes and Florian have tended to be relatively unobtrusive.

Sinda and Blasius

Sinda hatched in 1974 and was hand-reared by Sybille Kalas-Schäfer, along with her three sisters, Alma, Alfra, and Jule. The foster mother tried to develop an especially strong bond with Alma in order to observe closely her later behavior toward her offspring. This had the result of strongly bonding all the other sisters to the foster mother and to each other as

49

well. When Alma, Jule, and Sinda were later courted by the same ganders, Alma only rarely threatened Sinda. On three different occasions the latter responded by reverting to juvenile fluent cackling—in other words, treating Alma like her mother.

Blasius is one of the many still-living ganders that hatched in 1973 and were hand-reared by various coworkers. He belonged to the flock that Sybille Kalas-Schäfer looked after. Florian and Markus, along with a large number of other goslings, were led by Brigitte Dittami-Kirchmayer.

Background

The formation of the pair in 1975 was complicated by the bonding between the sisters. At first Jule seemed to be accompanying Markus and Sinda seemed to be associating with Blasius, but later Jule was often seen with Blasius and Sinda with Markus. This confused situation is apparently responsible for the implacable hatred that gradually developed between the two ganders. There were repeated aerial battles between Markus and Blasius. On one occasion, they landed near the Auingerhof, immediately went for each other with bent necks, and engaged in a battle with their wing shoulders in which Markus was beaten. In a later wing-shoulder battle, Markus emerged victorious and Blasius crumpled and fled. Markus triumph-called while approaching Sinda, who was still in a close-knit group with her sisters. The two ganders fought repeatedly over the course of the next two days. Blasius frequently had to flee to the nearby wildlife park, but he always returned after a brief absence. Markus stayed fairly close to Sinda, though avoiding her immediate vicinity. The hatred between the two ganders continued to grow, in 1975 leading to a dramatic aerial battle from which Markus emerged victorious. At a height of about fifteen meters, he struck Blasius so forcibly on the neck just above the shoulder that Blasius's wing was temporarily paralyzed and he plummeted like a stone. Luckily, he fell into a narrow but deep water-filled gulley. The wing paralysis lasted several hours and then disappeared completely.

Thereafter, Sinda associated closely with Markus, and Alma, Sinda, and Markus flew around together as a tight-knit group. But the hatred between Blasius and Markus persisted, and Blasius gradually became dominant. Sinda was once seen closely associated with Blasius. Shortly afterward, Alma, Sinda, and Markus flew off together and returned a few days later in close formation. Blasius returned a little later with Jule, having now formed a strong pair bond with her.

At the end of March 1976, Sinda was still accompanying her foster mother from time to time, and on one occasion they met up with Alma and Markus. Sinda immediately ran toward them and cackled with Markus. Before then, Sinda had often been seen with Florian and his mate Nat. Markus had by now become closely associated with Alma, although he continued to chase away any ganders that showed an interest in Sinda.

On April 1, 1976, Blasius was spotted alone, his wing coverts missing, and Jule was later found, decapitated, alongside an empty nest. Alma and Markus were nesting on a bank along the River Alm. We took away their clutch, because the area seemed dangerous, and they successfully hatched a second clutch of eggs.

History of the Relationship Between Sinda and Blasius

Sinda and Blasius lived as a stable pair from June 28, 1976, on. They did not breed successfully in 1977. Starting in the middle of April that year, Blasius was often seen alone or with other geese, usually not including Sinda. During the incubation pauses, both of them stopped off at the Auingerhof from time to time. On April 25, Sinda was heard giving loud distance calls while flying in from the direction of Lake Alm.

Florian replied but did not approach her. As she ate, Sinda continued to direct distance calls at the geese flying overhead. Blasius stayed at the Auingerhof. Florian approached Sinda with the bent-neck posture until he was three meters away, when she flew off in the direction of the Auingerhof, with Florian following. The next day, Sinda again flew in from the direction of Lake Alm. Florian flew up to her and accompanied her while showing the bent-neck posture. Blasius arrived alone at the Auingerhof after Sinda had left. Somewhat later, having taken Sinda away, Florian flew up from Lake Alm calling loudly. When Sinda arrived back at the Auingerhof on April 27, Blasius immediately went up to her. After she had eaten, the two went off together, but in a few minutes Blasius returned to the Auingerhof. Sinda was showing clear signs of incubation. The down feathers on her belly had been shed, she was very pale and thin, and her beak was frayed and scaly. But she apparently abandoned her nest a week later, and we were never able to find it.

In 1978, Sinda and Blasius bred successfully on the "floating island." Five goslings were hatched, and three of them made it through to fledging. In 1979, they hatched four goslings, all of which they brought back to Oberganslbach and reared to fledging. The following spring, because of deep snow, none of the geese except Sinda and another female incubated eggs. Sinda hatched out four goslings, which, as in the previous year, she reared to fledging at Oberganslbach.

During the breeding season of 1981, the harmony of this successful pair was disrupted. On March 14, Sinda was spotted near her nest site with Florian, and Blasius was nowhere to be seen. There are two likely explanations for the absence of a gander when he is supposed to be guarding the nest: either he is dead or he has been so badly beaten in a fight with a rival that his bond to his mate and to the nest site has been temporarily broken. We do not know what happened next, but ten days later Sinda turned up with Blasius at the Auingerhof, several kilometers from the nest site. Blasius was blocking Sinda in a clearly anxious fashion, and Florian looked as though he had taken a terrible beating. At this point, a third gander intervened. This was Ado, a male that had conquered the gander Gurnemanz and taken away his mate, Selma, in the breeding season of 1976. It was therefore unclear whether the disheveled Florian had been beaten by Blasius or by Ado. In any event, two days after being seen at the Auingerhof with Blasius, Sinda was sighted with Ado near the boathouse on Lake Alm, a few hundred meters from her nest on the "floating island." Again Blasius was nowhere to be seen. For the next few days, Sinda remained on the "floating island" incubating, with Ado keeping guard. Whenever Blasius tried to approach, Ado drove him away by flying at him savagely. Ado was remarkably tame and friendly, even responding to the very hoarse calls of the observer (who had a cold). Blasius remained nearby, although Ado repeatedly chased him away. On April 8, Sinda and Ado were seen together for the last time during the incubation pause, and Ado chased off Blasius, who made no attempt to fight. Three weeks later, Sinda and Blasius were together as if nothing had happened, and Ado was never seen again. From several cases we have recorded, we believe that ganders may withdraw from very tense situations by simply flying off.

Following Ado's disappearance, Sinda incubated six eggs— four of which were good—while Blasius stood guard. Two goslings hatched, and one of them was subsequently raised to fledging at Oberganslbach. In 1982, Sinda hatched six goslings from seven eggs on the "floating island." The family arrived at Oberganslbach with all six goslings, but it was scared away by woodchopping and moved downstream to the Auingerhof. The next afternoon, Blasius and Sinda reappeared at Oberganslbach with only two goslings; they had lost four during the return journey upstream.

At the end of February 1983, these two surviving goslings were still accompanying their parents. In the middle of March, another goose, Leni, hollowed out a nest cup on the "floating island." Although Sinda's old nest was left untouched, a territorial battle ensued, with Blasius launching aerial attacks on Leni's gander, Selmasohn. On March 18, Sinda was found sitting on five eggs in Leni's nest. Two days later, however, she was seen sitting on two eggs in her old nest—that is, in exactly the same place as in earlier years—and there were six eggs in Leni's nest. From March 22 on, Sinda regularly incubated three eggs, while the two ganders continued to fight for possession of the island. Whenever the two females took time off for an incubation break, the two ganders went for each other. Blasius lost a number of wing-shoulder fights, although Sinda actively intervened to support him. Finally, during a spectacular aerial battle, Blasius managed to grab his opponent in the air and pull him down to the water. He ducked Selmasohn in the water several times before his opponent was able to escape. Selmasohn put up no further resistance after that, leaving Blasius in full control of the island. He had launched his aerial attacks on Selmasohn from more than a hundred meters away—a distance that signified a very high level of aggression. Sinda hatched out three goslings, all of which were reared to fledging at Obergansl-bach.

In April of the same year, the Sinda-Blasius family met up with Claire (one of Sinda's daughters, hatched in 1979), her gander, Kasimir, and their goslings, and the goslings of the two families intermingled. The confrontation between Sinda and Claire was unusual, with the latter producing high-pitched cackling, a sign that she still recognized Sinda as her mother. A few hours later, when the families separated, all the goslings followed correctly.

At the beginning of March 1984, after the goslings from the previous year had separated from their families, Blasius was violently attacked by a gander pair and chased away. Sinda left her old nest on the "floating island" untouched and nested instead right alongside Leni's old nest. On March 10, she was found sitting on seven eggs, three of which were removed for hand-rearing. Three goslings hatched from the remaining four eggs, and as usual Sinda and Blasius moved on to Oberganslbach to raise them to fledging. On May 1, Sinda violently attacked an old, high-ranking gander with which Blasius had often fought. She pulled out some of his feathers, bit him, and chased him off with a blow from her wing shoulder. The gander's mate and Blasius both ignored the fight. The gander soon returned to attack the Blasius-Sinda family, which immediately swam away. During the altercation, one of their goslings became mixed in with the opponent's flock, but the next time the families met up, the stray gosling returned to its correct family.

In 1985, Sinda and Leni both incubated on the "floating island" again. Blasius left Leni alone, but he would not tolerate having Selmasohn around. We removed three of Sinda's six eggs for another goose. Although the three remaining eggs hatched successfully, the family arrived at Oberganslbach with only one gosling, and this one was lost within a few days. Sinda nevertheless continued to behave as if she were leading goslings. One of her daughters from 1983—the one that had associated loosely with the family the previous year—again began to accompany her parents, along with one of their offspring from 1984. Toward the end of the year, they were often seen on Lake Alm separated from the flock. Starting in October, they became particularly cautious in their behavior, and they occupied a very low rank in the flock.

In 1986, while Sinda was incubating at her old nest site, Blasius showed relatively little nest-guarding behavior and was seen alternately on Lake Alm and in the wildlife park. He seemed to notice his new family properly only when monitoring the nest after all four goslings had hatched. After its arrival at Oberganslbach, the family obviously moved up in the hierarchy. In July, there was a fight between Blasius and Florian, with Florian fleeing.

For the sake of simplicity, this account has omitted several less important episodes. Overall, with respect to the rearing of offspring, Sinda and Blasius were our most successful pair in the years from 1978 to 1986. Although we took eggs away from them on a number of occasions, they raised twenty-four goslings to fledging. At the time of writing, however, only one of Sinda's daughters—Claire, now eight years old—has reared her own offspring. If these findings apply to the other reproductively mature geese in our flock, it is easy to see why the population is not climbing sky-high.

Finally, let me say something about the special qualities, the individual characters, of these two geese. It is notable that Blasius almost always lost his first fights with rivals. That he emerged victorious in the end is possibly due to his special skill in aerial fights. Sinda has always been active in conflicts with other members of the flock. Blasius has generally preferred to stay in the background, but he is prepared to intervene in defense of his family when it is threatened, as long as the opponent does not appear invincible. When the family arrives at the feeding area, Sinda usually goes ahead while threatening or showing demonstrative vigilance, with Blasius and the offspring following. Sinda and Blasius generally occupy a middle position in the hierarchy among the families, and outside the breeding season their rank tends to decline.

Ada's Story

Ada's turbulent life history is a far cry from the typical biography of a greylag goose. Her egg was taken from a nest of wild greylag geese in 1949 or 1950. The egg was hatched by a domestic goose belonging to my friend Rolf Ismer, and the same goose raised her. There are a number of indications that she had contact with human caretakers while still a small gosling. As an adult bird, she was given to a Doctor Bauer in Oberkassel on the Rhine as part of an exchange, along with a three-quarters-wild gander with which she had presumably paired off. But the gander strayed away while out flying, leaving Ada as the only greylag goose in Doctor Bauer's waterfowl collection. In 1952, he donated her to our institute in Buldern, Westphalia.

Ada's behavior was more like that of a widow than of an unmated goose. She was very shy at first, partly because we clipped the primaries on one of her wings as a safety measure while she was becoming accustomed to Buldern. She did not join up with any of our young greylag geese that had just reached sexual maturity, but remained shy and tended to seek human company. She was frequently chased by the other geese, and the hand-reared "giants," Fasold and Fafnir, which were one-quarter wild, were especially aggressive toward her.

As Ada gradually settled in and became more active, she began to attack Röschen and Verena, two sisters that had paired up with the gander Adolar. Adolar and Ada often had violent confrontations with each other. They once engaged in an out-and-out duel with their wing shoulders, something that is very rare between a male and a female greylag goose. That battle ended inconclusively, but Ada was beaten later by another gander, Syrrhaptes. Ada showed a somewhat masculinized behavior in other respects, possibly because she had never lived with higher-ranking members of her own species in Oberkassel.

In February, Fasold began to approach Ada while neck dipping, which is an invitation to copulate. She responded, but not according to form. Unlike the typical female greylag goose, she simply turned toward the gander and swam straight after him. This unexpected immediate compliance obviously sent Fasold an ambiguous message. Not knowing what he was supposed to do next, he showed contradictory behavior. When the two were on dry land he chased her away, but when they were on the water he swam up to her and performed infatuated neck dipping.

Then another female—three-quarters domestic, like Fasold, and called Pummelchen because of her dumpy appearance—began to follow him. Domestic geese reach sexual maturity earlier and are more sexually active than pure wild geese. Fasold dipped his neck in the water with his head pointing toward her and repeatedly mounted her. Ada responded by intensifying her own courtship of Fasold. She swam just behind him, moving closer whenever Pummelchen tried to approach. Fasold and Ada were never seen to perform the triumph ceremony together, and only once did I see Fasold walking alongside Ada, far from the other geese, with a pronounced arched neck. She had clearly fallen in love with Fasold and remained constantly near him. He, for his part, gradually began to follow her, although Pummelchen usually followed openly behind him, showing hints of a bent-neck posture whenever he made this display to Ada.

In Ada's confrontations with Syrrhaptes, Fasold actively intervened twice, blocking the other gander's way when he tried to attack Ada. On February 19, Fasold was attacked by his brother Fafnir, and Ada attacked Fafnir with a mixture of aggression and fearful inhibition. She strode toward him until their bodies were almost touching and then froze with her wing shoulder raised to strike and her neck in an extreme elephant posture. She behaved exactly the same during a later encounter between the two "giants," but on neither occasion was there actual physical contact.

As spring approached, I had the impression that Ada's love for Fasold was waning, and in May I suddenly realized that she was following Fafnir. Because of the great likeness between the two brothers, I had failed to notice this at first. Strangely, Pummelchen had made the same switch in partners, and that also had kept me from realizing what was going on. It is possible that a fight had taken place between the two "giants," with Fafnir emerging the winner. Whatever the case, Ada did not show triumph-ceremony bonding with either brother in 1952, and there is no indication of copulation having taken place.

In December of that year, a male white-fronted goose

called Adus, who was unable to fly, began to court Ada intensively. He followed her with his neck arched, and in the water he swam parallel to her a few meters away while showing an extreme frigate posture. After making a series of both serious and sham attacks on other inhabitants of the pond, Adus directed intensive cackling (the final phase of the triumph ceremony) at Ada. Although she did not respond to it, she tolerated the presence of the small gander nearby. Also, she actively sought contact with humans. Not discouraged, Adus followed her persistently, and before long Ada began to join in his triumph-calling. It should be said at once that he remained faithful to her until 1962—by no means an automatic outcome.

During the winter and early spring of 1953, Ada showed a strong preference for associating with human beings. She especially liked to follow me, even to places that were difficult to reach on the wing, although she never came right up to me. I often went fishing for water fleas (*Daphnia*) in a broad ditch running through the middle of a willow wood. In order to land there, Ada had to drop almost straight down between the branches, brushing the twigs and flapping her wings. She skillfully carried out this flight maneuver, which is difficult and unpleasant for a goose, despite the fact that she was conspicuously disinclined to fly at all. Anatids that have once had their wings clipped are known to fly less than intact individuals do. That is why people who keep ducks or geese pinion them in the first year if they intend to leave them free to fly later on.

Ada's approaches to me were sexually motivated, for she responded to me with the intensive neck dipping that is part of the prelude to mating. The way I dipped my water-flea net rhythmically into the ditch, in my search for food for my fish, may have acted as a releasing stimulus. That was the only time we ever observed a female goose performing toward a human the instinctive motor patterns that derive from the copulatory context.

Adus suffered greatly from Ada's excursions, which often

led her away from the pond to areas frequented by humans, but he would follow her as far as he could, given his inability to fly. As soon as she settled on the water, he would direct intensive neck dipping toward her and would shower her with triumph-calling invitations, to which she made little response. We never saw a clear response to his neck dipping, and she certainly did not allow him to mount her.

Nevertheless, on March 22 Ada was observed seeking out a nest site on a platform we had built for sunbathing. She began to build a nest with material that she removed from the screen of reeds. When I discovered that she had already laid an egg, I took the risk of moving the nest and the egg, under her gaze, to a more suitable site just a few meters away and behind a protective fence. To my delight, she simply followed me and immediately sat on the nest, which I tried to make more attractive by adding some straw and hay. I must confess that I had an overwhelming feeling of achievement at having induced a greylag goose to nest at a site other than the one she herself had selected.

When Ada laid her fourth egg, on March 30, and was firmly installed on the nest, Adus began to stand close by on nest guard. In response to her nest call, he would attack anything that approached. Ada was extremely thorough in her incubation, taking only the briefest of pauses, which never amounted to more than ten minutes in any one day. Halfway through the incubation period, the clutch was found to be unfertilized, and I removed the eggs from the nest. Ada at once became very shy, avoiding other geese and usually keeping to the edge of the flock. Adus continued to follow closely behind her.

In the autumn, Ada began to seek contact with the domestic geese in the farmyard of nearby Buldern Castle, where Adus, unable to fly, could not follow. As soon as she flew off, he would swim or run incessantly to and fro uttering distance calls. Ada paid less and less attention to him— although I was a little to blame, because I spent a lot of time talking to her and feeding her. She would follow me to places that the other geese never visited. For example, she often flew onto the roof of the hut beside the sunbathing platform to keep me company and would stare down at me for hours on end. Despite this strong attachment, however, she never became really tame and never of her own accord came closer to me than three meters.

When the ponds froze over during the very cold weather of January 1954, all the geese grew noticeably more timid; only Ada became increasingly tame toward me. We had to take special action to keep the last open area of the Buldern goose pond from freezing over. Using a hoe and a saw, we cut out a large chunk of the ice and pushed it under the surrounding layer of ice. We did this by having one heavy person (me) stand on the perimeter and a lighter person stand on the freed chunk, pushing it downward in the water and thrusting it under the surrounding ice by quickly stepping backward. With this somewhat dangerous procedure, we managed to clear an adequate surface area and were looking on exhausted as the geese began to gather. Then, for some reason, I stumbled and fell into the water, right in the middle of the geese. They all took off in fright and, because the fog was thick and night was falling, they did not land but flew off in tight formation. Only Ada and the geese of mixed descent remained behind. The pure greylag geese eventually returned in small groups, only a few of them disappearing for good.

At the beginning of March 1954, Ada resumed her excursions to visit the domestic geese and spent much of each day with them. She would fly back to our pond to sleep as night approached. Adus would greet her passionately every time, although she was by then spending only the early morning and late evening with him. Ada was regularly seen being courted and mounted by one of the domestic ganders in the farmyard. Despite my calling and tempting, she would never leave him during the daytime, and she also would not respond to courtship and triumph-calling invitations from Adus.

On March 16, 1954, we observed the domestic gander

mounting a domestic female in Ada's presence. Ada's response was a violent attack on the copulating pair, but when we called her she immediately took off and followed us back to the pond. We never saw her again with the domestic geese.

Ada now stayed put on our pond, constantly accompanied by Adus, although she had never responded to his triumph-calling. He continued his courtship unabated, displaying parallel swimming, an extreme frigate posture, and the spreading of his tail feathers, but not neck dipping. We had implanted some crystals of testosterone under the skin on his head in February, and this clearly made him more courageous, with the result that he climbed markedly higher in the social hierarchy of our goose flock.

In March, Ada again began to seek a nest site on the sunbathing platform, and, as in the previous year, she allowed herself to be talked into moving to an alternative site. During the breeding season, she responded more strongly to the courtship of the white-fronted gander. The two of them together drove off other geese, and they were frequently seen copulating. On March 26, Ada laid her first egg, and she had started to incubate to some extent by the time the third was laid. After laying another two eggs, she showed intensive incubation. As before, her incubation breaks were very short, never amounting to more than ten minutes per day.

As the incubation period progressed, Adus guarded the nest with ever-increasing intensity and proximity. He drove away all intruders, whether they were bigger or smaller than he was. He had one especially violent fight with a male Canada goose named One-eye that clearly outmatched him, during which Ada left the nest and attacked the Canada gander with her wing shoulders. Adus and Ada then performed a very long and intense triumph ceremony together.

Three enormous golden yellow goslings hatched from Ada's eggs on April 28—obviously the offspring of the domestic gander from the farmyard. The next day, after we had helped the goslings to descend, Ada left the platform.

Adus led the family in a very agitated state, behaving with extreme aggressiveness toward any geese that approached. He had many fights with higher-ranking opponents and returned to his family with loud triumph calls, in which Ada now regularly participated. The spring and summer of 1954 were the only happy times that Adus ever experienced.

In October, the cohesion of the family was somewhat weakened when Ada began to fly off on long trips with her three white offspring while Adus was forced to stay behind, calling loudly in agitation. In November, we sacrificed Ada's two male offspring, because we did not want to have too many crosses with domestic geese, but it had little effect on the cohesion of the family. Ada flew off as often as before, now accompanied only by her remaining daughter, whom we had named Adakind. Adus continued to be very aggressive, maintaining dominance over two pairs of white-fronted geese, a trio of bar-headed geese, and even the "quartet," a group of greylag geese that included three powerful ganders. Adakind was occasionally seen alone after January 1955, but when she gave a distance call both Ada and Adus rapidly joined her.

During a spell of mild weather in late January and early February 1955, Ada began to seek a nest site again. She repeatedly flew to the platform and performed transfer movements with the nest material there. She paid less and less attention to Adus and her daughter and barely responded to the triumph-calling of the white-fronted gander. She avoided any geese that Adus had recently defeated. All the geese took to attacking Ada and driving off Adus, whose aggressiveness had declined markedly. Apparently, a weakening in the bond between mates is directly linked to a simultaneous loss in fighting potential.

In his courtship behavior, Adus now directed his neck dipping more often toward Adakind than toward Ada. One time, when he directed his courtship clearly at Ada, Adakind pushed between them and displayed intensive neck dipping toward Adus. Over the next few days, his triumph-calling was sometimes more intensively directed at Adakind than at Ada. This kind of thing often happens when a mixed pair of geese, one domestic and one wild, raises offspring. Another gander that would later play a part in Ada's life was the son of a pure-bred male Canada goose and the three-quarters-domestic female Pummelchen. When, in the first spring after his birth, this hybrid gander became sexually motivated at the same time as his mother, the male Canada goose, showing no sign of readiness to mate yet, was not at all disturbed to have his son (appropriately called Oedipus) regularly copulate with his mother. A little later, when the Canada male himself became sexually motivated, the picture changed radically.

Starting in the middle of March, Adakind was seen more and more often with two male snow geese, and she allowed both of them to mount her indiscriminately. Afterward, she would perform the triumph ceremony with both of the snow geese and then would sometimes return directly to her parents with a triumph call. Now and then Adakind displayed neck dipping toward her mother, and on one occasion she climbed onto her mother and tried to mount her. We have seen such behavior only in geese crossed with the domestic form, and Adakind's precocious attainment of sexual maturity was probably also due to her mixed origins.

Toward the end of March, Ada abandoned her family altogether, leaving Adus swimming around alone and calling loudly after her. At this point, Ada was intensively courted by Oedipus, the Canada goose hybrid we have just mentioned. She responded in kind to his neck dipping, actively joined him in performing the triumph ceremony, and swam after him. Adus clearly saw all this, and whenever he saw Ada he would approach her with triumph calls, but she took absolutely no notice of him. Sometimes he swam after Oedipus and Ada, but if he made himself too obvious Oedipus would attack and drive him away. Yet this did not stop Adus from swimming after the pair. Oedipus and Ada often flew around together, however, and so they were often seen without Adus, although

Adakind and the two male snow geese sometimes followed them instead. One of the snow geese often courted Ada, and after Oedipus had mated with Ada, this male and Adakind often displayed triumph-calling. Adus remained as close to Ada as he could. He did not dare interrupt Ada and Oedipus when they were mating, but he often showed intensive courtship, rearing stiffly out of the water in an extreme frigate posture, his neck and chest raised high, while softly uttering the departure call. Following the pair's postcopulatory display, which was usually performed with great intensity, Oedipus often swam over to Adus and chased him away.

On March 26, Ada began to remain close to the platform, often flying onto it to prepare a nest hollow. Oedipus and Adus also stayed close by, although Oedipus would drive away the small male white-fronted goose whenever he directed courtship behavior at Ada. If Oedipus then returned to Ada while triumph-calling, she would avoid him or simply not respond. If he chased away other geese, however, Ada would actively join in his triumph-calling. Apart from this, she took absolutely no notice of Adus. When Ada laid her first egg, on March 27, Adus stood on the footbridge in an aroused state. But as soon as Adus uttered the nest call, Oedipus flew up and drove him away, although he otherwise avoided the platform. Ada was very agitated and scarcely left the nest.

On March 28, another suitor joined in. This was an old, aggressive male Canada goose with considerable fighting experience who had lost his former high rank after a fight with a rival and had therefore been christened the Underling. After chasing Oedipus and then Adus away several times, he was suddenly seen standing near the platform. He was producing soft, deep growling vocalizations and directing strange undulating movements of his neck at Ada's nest. Ada excitedly produced the nest call in response and pushed her head through the screen of reeds in the corner of the platform where she was incubating. In fact, an opening gradually wore away in the reeds where Ada and the Underling looked at each other. He responded to her nest calling with the rolling call and tried to jump up onto the platform. But the minute his head appeared over the edge, Ada viciously bit at him, further increasing his arousal. Attempts by ganders to "adopt" incubating females, along with the nest and the offspring, are not uncommon, and doubtless this was such a case.

Neither Oedipus nor Adus could see the Underling's intrusion from their guarding positions, and the calling repeatedly attracted them. The Underling managed to beat Oedipus and chase him away, but he tolerated Adus's presence, as if to say that this dwarf could not be taken seriously as a rival. Adus did not dare approach the nest, but stood guard on the footbridge some meters away. If the Underling turned his back, Adus would attack him from behind; but if the Canada goose turned around, he would retreat, acting as if nothing had happened. The Underling did not respond to Adus's attacks. Whenever Ada left the nest, the Underling would hurry after her and passionately regale her with triumph-calling, but he had been pinioned, and Ada was able to avoid him by simply flying off. She made no response to his triumph-calling.

Oedipus also appeared whenever Ada took a break from incubation. At first she responded only hesitantly to his triumph-calling, and none of her later responses were very strong, but the two geese would fly around together and Ada would graze alongside Oedipus. On one such occasion, Oedipus met up with the Underling, and they engaged in a violent wing-shoulder battle. Oedipus, emerging victorious, chased the Underling a considerable distance away. Afterward, he triumph-called intensively with Ada, and this time she participated fully. The Underling subsequently made several attempts to stand guard by Ada's nest, but Oedipus always chased him off. After his victory, Oedipus himself began to stand guard by the nest, although he was not as persistent at it as a pure-bred male Canada goose would have been. In greylag geese, such behavior is less well developed.

The defeated Underling continued to stay close to Ada's nest much of the time, and, strangely, a friendly relationship developed between him and the white-fronted goose Adus. A few days after the decisive fight, Adus and the Canada goose swam away from the platform together, triumph-calling in unison, with Adus adopting an extreme frigate posture and repeatedly uttering the departure call in front of the platform. Sometimes the two ganders were joined by the hand-reared greylag gander Schwarzblau, then a year old, and Adus directed weak courtship behavior toward him. Many years later, Schwarzblau was to become Ada's great love and successful mate.

On April 1, 1955, an extremely aggressive, high-ranking pair of Canada geese established a nest on the dam across which the path to the platform ran. Because they attacked any humans who used the path, we referred to them as the "bandits." The advanced stage of the breeding cycle this pair had reached did not prevent the gander from immediately taking up a position near the platform on which Ada's nest was located; he had spotted the incubating goose through the opening in the reed screen. Ada became very agitated, uttering the nest call and biting the intruder's beak when he peeked through. None of Ada's other suitors had dared come this close to her. By the next day the male bandit had lost interest in Ada, however, and displayed triumph-calling only with the female bandit. In defending their own nest site, the bandits drove all three of Ada's suitors away from their guarding places. Oedipus tried a few more times to display neck dipping to Ada, but then, becoming thoroughly confused, he swam to the nest of his aunt Sinchen, displayed neck dipping there, courted his mother, Pummelchen, as he had done before, and started swimming around with his sisters. It is typical for a goose that has lost its mate to return to its family, and Oedipus paid no further attention to Ada.

The Underling had plummeted in the rank order after his defeat by Oedipus. All the other geese chased him away, and he remained mostly on the periphery of the flock. Adus, meanwhile, was showing intensive courtship of the young male Schwarzblau—later renamed Adonis—with whom he had developed an intensive triumph ceremony and in the process had regained his former aggressiveness. Neither male took any notice of Ada. After April 3, she concentrated on her incubation, taking the briefest of pauses and rarely uttering nest calls. Only if she had a conflict with another goose during an incubation pause did Oedipus or the Underling join in her nest calling, although they never ventured into the danger zone.

When Ada had been incubating for a week without any gander standing guard, some unknown disturbance frightened her off the nest, causing her to fly straight upward. This provoked general excitation among the other geese. She herself uttered only a few nest calls, but Adus abruptly went after her, intensively triumph-calling and meeting her with a greeting ceremony. Although Ada did not respond, Adus swam after her, grazed alongside her, and followed her when she returned to the nest about ten minutes later. From then on he faithfully stood watch by the nest, though not with the same dedication of the previous year. Now and then he swam away and spent some time with Schwarzblau before returning to the platform. Because Ada's eggs turned out to be unfertilized, we replaced them with some fertilized eggs.

During one incubation pause toward the end of the breeding season, Ada chased away some snow geese. At this, Adus hurried over and gave triumph calls; aroused by the fight, Ada joined in for the first time. Then, shortly before the hatching, Ada got up from the nest and ran to and fro uttering loud nest calls. The Underling, Adus, and the male bandit all appeared immediately. The Underling thrashed Adus several times, and the bandit thrashed the Underling. After the fighting was over, the Underling timidly returned to the platform but intervened when Adus tried to do the same.

When Ada's substitute goslings hatched, all her suitors—Oedipus, Adus, and both of the male Canada geese—turned up and triumph-called together. Wild, indiscriminate fights then broke out among the ganders, and we thought it best to remove Ada and her goslings to a safer place. But she could not settle down in the aviary where we placed her—the many triumph-calling invitations were making her nervous.

At this point, yet another male Canada goose—Blau-Rot, later called Adamann—joined Ada's "harem." He led the family when Ada was released from the aviary with her goslings, and she soon began to join in his soft triumph-calling. Adus faithfully followed the family, which tolerated his occasional attempts to lead the goslings. Soon both ganders were leading the family, and over the course of the winter the family developed a communal triumph ceremony.

In December 1955, when all the geese in Buldern were laboriously and patiently captured and transferred to Seewiesen, Ada was left behind with a handsome greylag gander called Oswald. But my hope that a bond would develop between the two was not fulfilled. Although they went around together, they showed no inclination to perform the triumph ceremony. Oswald did not show courtship of Ada, and the coworker who was keeping the records added this acid comment to her notes: "Why should he?" Toward the end of February 1956, Ada began flying regularly to spend the day with the domestic geese in the farmyard, where she stayed close to the domestic gander.

All the transferred geese had the primary feathers on one wing clipped before they were let free on Lake Ess, and in these unfamiliar surroundings they kept close together and soon developed fixed habits. They would spend the day near the institute building on the eastern shore of the lake and pass the night on the floating moor on the western shore, opposite the institute.

On March 21, 1956, Ada and Oswald were captured and transported to Seewiesen. Ada, her wings unclipped, was placed in a small quarantine cage. It was quite dark when she arrived, but within a few hours Adus was standing beside the cage and intensively triumph-calling. He must have recognized Ada's voice. A few days later the male Canada goose Blau-Rot also turned up at Ada's cage. Although he had not once triumph-called with Ada's goslings or with Adus in the interim, he promptly chased Adus away and triumph-called toward Ada. When she was released, both ganders followed her displaying neck dipping. By the end of March, Ada and Blau-Rot were behaving like a firmly bonded pair, triumph-calling, neck dipping, and engaging in copulation. Adus was driven away by the Canada goose, but for a while he followed the pair, displaying the frigate posture and courting Ada from a distance of ten meters. Later he was often seen with Aida, a sister of Oswald and Adonis (the former Schwarzblau), and eventually he began to court her.

When Ada began to lay, in the first days of April 1956, the two ganders remained close to the nest only for a short time. Then Blau-Rot, or Adamann, began to court the female Canada goose Jolanthe. Adus swam with Adamann; but neither of them stood guard by her nest. Not until Ada's two goslings hatched did Adus and Adamann return to the nest, and then all three adult birds performed the triumph ceremony together.

Late the next evening, there was a violent fight between Ada and Jorinde, a female Canada goose. Jorinde had tried to kidnap Ada's goslings, which had, in fact, followed her and her gander, Tristan, for a short distance. Ada, Adus, Adamann, and Jolanthe all intervened in the confrontation, and so did the "bandits." Ada was left afterward with only one gosling. The other was seen with the "bandits" and also with the pair Tristan and Jorinde, who were leading two offspring of their own.

In February 1957, Ada, Adus, and Adamann were still closely associated. Then, toward the end of March, while they were engaged in a particularly rousing performance of the triumph ceremony, Ada abruptly turned on Adus and thrashed him, although she later triumph-called with him again. At the beginning of April, when Ada was incubating, the male snow goose Grün tried to rape her on the nest while Adamann was away feeding. Grün climbed on the nest and grabbed her by the neck, but she attacked him and beat him with vigorous blows of her wing shoulders. When he did not let go, she attacked with ever-increasing ferocity. Just then Adamann hurried up and helped Ada get a firm hold on Grün, and both of them struck him repeatedly until he lay completely exhausted on the water. Scarcely had Ada released her grip, however, then he pulled himself together and rushed back into the fray. But after one more sound thrashing, he fled.

In the spring of 1958, Ada was back with Adus again, although he had directed his attention the previous summer toward the greylag goose Gesine, who was paired with the male snow goose Schneerot. Schneerot and another male snow goose, Schneeblau, now courted Ada, and, paradoxically, Schneerot stood guard by Ada's nest rather than by his own mate's. Ada allowed him to mount her, and afterward she uttered a call that we have identified in the postcopulatory display of the greylag gander. If Adus also tried to stand guard by the nest, there was friction with Schneerot, but when the two ganders grazed, or were away from the nest for another reason, they stayed together. Adamann was now fading into the background.

In the first days of May, Adonis, the gander who had previously displayed courtship behavior toward Adus, was seen almost constantly with Ada, Adus, and the male snow goose. All three ganders displayed neck dipping toward Ada, although Adonis's was the most intensive. He was very aggressive, often chasing other geese away, and repeatedly attempted to direct his triumph-calling at Ada. But she avoided him for now.

Then, early on the morning of May 14, 1958, I saw

courtship taking place between Ada and Adonis of an intensity that I had never before witnessed in greylag geese. The two geese, oriented parallel to each other and about five meters apart, were walking extremely slowly in the manner described by Heinroth. Adonis alternately displayed an extreme bent-neck posture and an equally extreme form of demonstrative vigilance (a pattern that I also refer to as the polite-alarm display). It was then that I discovered the female counterpart to the bent-neck display of the gander. Ada's neck was held stiffly upward, at a more pronounced angle to the shoulders than in the male, and the head was turned down more sharply. The head was tucked in and pressed tight against the lower part of the neck, and the feathers on the back of the neck were somewhat fluffed, just as in the male's bent-neck posture but also resembling the submissive-neck posture.

The two geese went into the water after a few minutes, and both immediately adopted the frigate posture. Ada showed this in an extreme form that I had not seen before in a female goose. She immediately displayed neck dipping with equal intensity, and copulation abruptly took place, followed by an absolutely fantastic postcopulatory display. I can express the fire of the entire sequence only by repeating that I had never seen mating behavior of such intensity in greylag geese. Ada and Adonis remained close together after that.

A remarkable scene is recorded in the notes for August 10 of that year. Adus walked past Ada with a bent-neck posture, but then he stopped and began to preen himself, as if experiencing a powerful motivational conflict. Adonis approached, but instead of attacking Adus he chased away a duck and ran back to Ada with his neck stretched forward. No triumph-calling followed. Ada and Adonis simply began to preen themselves, as Adus was doing some two meters away. As of the autumn of that year, there is no further record of Adus interacting with Ada.

Adonis and Ada now formed a firmly bonded pair that

occupied a fairly high rank in the hierarchy. In 1959, Ada laid three eggs and raised two goslings, one of which was later lost. The other gosling stayed with the family until February 1960. In that year, Ada again had two offspring, and the family formed a loose association with other geese that were leading offspring. In 1961, Ada hatched four goslings from five eggs; the fifth egg was bad. One of the goslings turned up missing the day after hatching, and by June 1 the pair had only one gosling left. The following year, they were able to raise only one of the three goslings that were hatched. In February 1963, this surviving goose was still living with her parents, along with one of Ada's sons from 1960.

On April 9, 1963, Ada swam up to her nest with a piece of straw in her beak. She stayed in the water with her eyes closed and seemed exhausted. Later in the day, she was seen staggering around the meadow dragging her wings. We took her to the veterinarian, who diagnosed her as egg-bound and crushed the egg trapped in her belly. But Ada failed to recover. She was found dead the next morning beside the entrance to her nest.

Adonis did not spend any time in mourning, but immediately began to court other geese. It was not until August 1964, however, that he was seen with a goose with which he later formed a firm pair bond.

The "Quartet"

Max, Kopfschlitz, and Odysseus all hatched in 1952 and were hand-reared. As often happens with hand-reared brothers, they developed an intensive family triumph ceremony, which smoothed the way for a strong bond between the three brothers. A fourth brother, Moritz, was driven away by Max, and their sister, Schiefschwanz, gradually loosened her attachment to the group and eventually became independent.

It was apparent from the first years of the institute in Buldern that the bonding of ganders is not influenced by the incest taboo or by any other restrictions on pair formation.

The three ganders were joined by a female goose called Martina (not the original Martina), who had also been hand-reared but came from another group of goslings. These four geese remained together most of the time and came to be known as the "quartet." Sexual behavior in the form of neck dipping was at first seen only between Max and Odysseus. The neck dipping either led nowhere or would be followed by the two ganders trying to mount each other, in which case Odysseus would try to bite Max and chase him away. In the resulting excitation, Kopfschlitz and Martina would usually come running, and all four geese would then engage in triumph-calling. Martina occasionally tried to display neck dipping toward Max. If he took any notice of her, however, Kopfschlitz would push between them and chase her away. He did the same whenever Max tried to direct triumph-calling toward Martina. Odysseus and Kopfschlitz, like Martina, actively courted Max, who was clearly the central figure in the quartet.

Nevertheless, the bond with Max did not prevent Odysseus from maintaining a mating relationship with a one-quarter-domestic goose named Sinchen. He used to meet her, as if by arrangement, near the outlet of the lake. He often flew in from some distance away, and the two would quietly display neck dipping and then copulate. Afterward, however, Odysseus would fly straight across the lake to rejoin Max and Kopfschlitz and would perform an extreme form of the postcopulatory display toward Max.

The partnership between Max and Odysseus ended dramatically. On March 17, 1956, I heard a loud crash from the direction of the aviary near the goose house. I rushed over and found Max and Odysseus engaged in a vicious fight with their wing shoulders, which continued for some time. A fight with wing shoulders usually lasts no longer than a few seconds, and I had never witnessed one like this.

Expressions of "love" between ganders can, in fact, sometimes take a form that looks suspiciously like fighting. The typical change in orientation toward the head of the partner becomes more and more sharply angled, until the partners come face to face in exactly the same way as two ganders that are threatening each other, although they utter loud rolling calls rather than cackling. But this threat-laden situation can at any moment give way to pressed cackling, a behavior pattern that reflects extremely strong bonding. Jürgen Nicolai has observed with various birds, most notably bullfinches, that ritualized bonding patterns can lose their ritualized components at a very high intensity. Motor patterns that are derived from aggressive behavior can abruptly switch to the original, unritualized form of fighting with the beak. Something similar occasionally happens with the triumph ceremony of geese. In their excitement, two friendly geese may suddenly find themselves angrily confronting each other. Any fight that ensues will be far more intensive than any other fights.

Odysseus left the "quartet" after this fight. He acted tamer over the next few days and sought human contact. In addition, the very next day he directed a full triumph call at Sinchen, evoking an almost rapturous response. From then on, Odysseus and Sinchen formed a definite pair. He avoided the other members of the "quartet" and occasionally was vigorously chased by them.

Max now directed his sexual behavior more clearly at Martina, evoking a typical jealous response from Kopfschlitz. Kopfschlitz guarded Max and generally tried to keep him away from other geese. Toward the end of March, Odysseus made several feeble attempts to rejoin the "quartet," but Max and Kopfschlitz (especially the latter) repeatedly drove him away. Max directed his triumph-calling more and more at Martina, and this irritated Kopfschlitz. On March 23, the two ganders had a brief fight, but they swam together afterward and tried to mount each other later in the day.

When the time came to seek a nest site, at the end of March, the relationship between Max and Kopfschlitz became particularly close. Martina, on the other hand, spent much of her time apart from them. The two ganders often exchanged intensive neck-dipping displays and tried to mount each other. On one occasion, a loud triumph ceremony involving Max and Kopfschlitz gave way to clearly aggressive threats, with the body held low and the neck stretched out in front. There was no actual fight, however, and afterward the two ganders were seen standing peacefully together.

Kopfschlitz was once observed directing his triumph-calling toward a human, with his neck stretched far forward. Max directed his triumph-calling toward Martina with steadily increasing frequency, and he also mounted her. During nest-seeking behavior, however, the ganders were regularly seen without Martina, and she was obviously looking for a nest alone. Given the unreliability of Max and Kopfschlitz, that was undoubtedly a good strategy. During this period, Martina became very shy. In May, the two ganders were seen looking for a nest together. But it was only in the middle of May that we found Martina in a breeding box, her feet pale and her belly slim. Presumably she was incubating. Max and Kopf-schlitz valiantly defended the nest, and the three remaining members of the "quartet" were constantly together during this time. They had frequent conflicts with a pair of Canada geese that had established a territory right alongside. In these fights, Max proved to be more aggressive than ever. Unfortunately, Martina's clutch was unsuccessful. Odysseus and Sinchen were often seen near the three geese, but they were usually chased away.

In February 1957, with spring approaching, Kopfschlitz, Max, and Martina stayed close together, although Martina was sometimes seen alone. Kopfschlitz, having previously directed triumph-calling at humans, now began to attack them. Max continued to direct his triumph-calling more strongly toward Martina than toward Kopfschlitz, while the latter usually directed his triumph-calling at Max. Sometimes, right in the middle of a cackling sequence, Max would switch his orientation away from Martina and toward Kopfschlitz.

Max and Kopfschlitz spent one afternoon wandering through the bushes, clearly looking for a nest site, and they also flew together across the lake to search on the other side. This year, too, a wing-shoulder fight broke out between Max and Kopfschlitz when they were trying to mount each other. But afterward they swam past each other showing the so-called cut-off behavior. Following one intensive aerial attack by Kopfschlitz on an unfamiliar goose, the two ganders displayed triumph-calling together, and then Max ran to Martina and tried to repeat the behavior with her. But Kopfschlitz rushed ahead of him and tried to divert the triumph calls by edging him to one side and shielding him. Only once were the two ganders seen together directing triumph-calling at Martina, and then Kopfschlitz was clearly doing so with far less intensity.

Toward the end of March, I again saw the two ganders more frequently without Martina. They often flew off in unison or would spend the night together. Martina would not accompany them, even when they went far off into terrain with good cover, as is usual during the nest-seeking period. Although she remained behind, she showed through her distance calls that she really would have liked to go nest seeking with them. Max would perform intimate triumph-calling with Kopfschlitz, but whenever the latter was not around, Max would turn to Martina. As in the previous year, Martina went off to seek a nest alone. On one occasion, when Ada and her family attacked Martina in front of a nesting box, Kopfschlitz and Max came up and drove the attackers away—the first time they had actively supported Martina.

Kopfschlitz now became the more aggressive of the two ganders and began to display triumph-calling more frequently. He was also observed swimming after Martina and directing his triumph-calling at her, though she swam fearfully away. On one occasion, when Max was mounting Kopfschlitz with no more than token resistance, Martina swam close alongside and began to bathe, as if she herself were being mounted. Max, obviously strongly motivated for breeding, performed transferring movements in the water as if nest building.

The two ganders often attacked other geese and chased them a long way. Sometimes, after such an attack, they would return to Martina but would walk right past her without greeting. Martina sometimes followed them and joined softly in their triumph-calling. Max occasionally mounted Martina, but would perform the postcopulatory display alternately with Kopfschlitz and with her. Then the two ganders would swim to shore together, leaving Martina alone on the water.

By April 11, Martina had laid her clutch, and she managed to defend her nest, by means of aerial attacks, with no help from the ganders. Throughout the breeding season, she was far more aggressive than other females without ganders. As Martina began to incubate the eggs, Max and Kopfschlitz intervened a few times to defend the nest, but afterward they usually performed the triumph-calling together, not with Martina. On the rare occasions when they directed their triumph-calling toward Martina, she sometimes failed to respond. A week later, Martina had four eggs. The two ganders now stood guard over the nest and defended it. They also began to direct their triumph-calling toward Martina and showed great agitation whenever she left the nest. But Martina had stopped following the ganders and paid absolutely no attention to them. Despite their careful guarding of the nest, the two ganders also flew around together a good deal of the time. They would attack any human who was checking the nests and then would display intensive triumph-calling together and try to mount each other, after which the gander underneath would swim away and the gander on top would drift some distance in the water.

On April 26, 1957, Max displayed such vigorous triumph-calling toward Martina that he bit her on the neck—a sign of maximum intensity. At a later date, Max and Kopfschlitz were seen swimming face to face, as they always did when they were about to attempt to mount each other, but this time they struck out with their wings instead. A violent fight with wing shoulders ensued, and afterward the two ganders returned to land and stood some distance apart as if in embarrassment. Then they approached each other and performed intensive triumph-calling. At this point I intervened by chasing Martina away from her nest, and the two ganders swam over and directed their triumph-calling toward her, with Max's calling the more vigorous. Kopfschlitz now approached Max and began biting him and then Martina on the neck in turn, showing greater aggression toward Martina. Kopfschlitz had apparently been the victor in the preceding fight. A week later, Martina hatched three goslings. The ganders stayed close to the family and defended the goslings very effectively.

In March 1958, the situation had changed slightly in that Martina now occupied a somewhat higher rank. The offspring from the previous year were still in the family, and the two ganders usually led the group, with Martina following behind. Max was now directing his triumph-calling at Martina as well as at Kopfschlitz, while the latter directed his triumph calls only at Max. When the two ganders occasionally tried to mount each other, the encounter was usually marked by tension, especially if they were face to face and attempting to mount from the front. Max repeatedly mounted Martina, with Kopfschlitz participating in the postcopulatory display. One time, Max first tried to mount Kopfschlitz and then mounted Martina instead. On another occasion, all three geese displayed neck dipping together, and then Max mounted Martina and performed the postcopulatory display with her. The same day, Odysseus unexpectedly attacked Max and Kopfschlitz, and both avoided him.

From this point on, Max directed his triumph-calling more strongly toward Martina, even when Kopfschlitz approached and tried to call with him. Max chased off many other geese and in general became very aggressive. On March 25, one of the ganders was seen mounting Martina. We suspected it was Kopfschlitz, for the mounting was slow and hesitant. Afterward, the second gander approached, and all three performed the postcopulatory display and bathed. The offspring from the previous year were still with the family, and once they mobbed Max while he was mating with Martina. The same day, Odysseus attacked Max and chased him away, and then drove away another goose as well.

Starting toward the end of March, Max and Kopfschlitz would repeatedly fly up to Odysseus, and Kopfschlitz would attack him savagely. Odysseus continued to be chased by Max and Kopfschlitz through most of April. Apparently the nests of the two families were close to each other, since the attacks always took place when the ganders were nest guarding.

By the end of April, Max was directing his attention almost exclusively toward Martina. On April 28, there was a violent wing-shoulder fight between Max and Kopfschlitz. Other geese rushed up, mobbing, but Max drove them off once he had beaten Kopfschlitz. Kopfschlitz, very weak, threatened the other geese. Then he staggered to the shore, and while he lay there with his neck stretched out in the submissive posture, Max attacked him again. We know of only a few cases like this, in which the defeated rival's submissive posture has failed to stop the victor from launching a further attack. Max eventually pulled away from Kopfschlitz, snapped at him a number of times, and then returned to stand in front of Martina's nest, where he shook his wings and resumed his guard. Kopfschlitz dragged himself weakly to the shore and crept into the bushes.

The next day, Kopfschlitz stood around in a very huddled posture uttering low, single-syllable calls in a rhythm resembling that of lost piping. He frequently shook his beak and preened himself. Later on, he would swim close behind Max— much closer than before the fight. But by the end of the breeding season the "quartet" was reunited, and they raised a gosling together. In the autumn, Max and Kopfschlitz again performed the triumph ceremony together, just as intensively as they had in the spring. They seemed to have declined in rank, however, because now when they were attacked they gave way more frequently.

From then on, the relationships between the members of the "quartet" became increasingly confused, especially because the offspring raised by Martina were now included in the triumph-calling party. Kopfschlitz disappeared on March 28, 1962 (probably taken by a fox while looking for a nest site). Max displayed intense grief for a time. He sought contact with humans, stayed away from his family for a number of days, and adopted the so-called sneaky posture. After April 2, he was often seen with a young greylag goose, with whom he stayed until the middle of the month. Then he started courting another young goose. But she was also being courted by another gander and ended up choosing him. Over the next few years, Max formed a variety of associations with other geese, both male and female. In March 1966, he began to stay with one certain female goose and twice raised offspring with her. In March 1968, he disappeared and was never seen again.

Theoretical Aspects

It did not seem a good idea to overwhelm my readers with a flood of theoretical information before I had provided some examples of greylag geese and their activities. Having done so, I shall move on to catalog the behavior patterns in the repertoire of the greylag goose, which, fortunately, are not unlimited in number. The systematic description of all these behavior patterns is known as an *ethogram*, although the term more accurately refers to the species-specific instinctive patterns.

The ethogram will give us a complete list of all the different kinds of behavior that the bird can show, and this considerably simplifies the investigation of its behavior. By contrast, in mammals, particularly the highly evolved primates, motor patterns modified by conditioning play a significant part. Accordingly, there is no hope of compiling a complete catalog of all the behavior patterns that it is possible for an individual mammal to perform. The much simpler structure of the behavioral system of the greylag goose gives us a great advantage. We can give a name to every observed behavior pattern, knowing that our worst problem may be that a pattern represents a mixture of several recognizable components. In other words, the ethogram constitutes an almost complete list of everything a greylag goose will do.

The Instinctive Motor Pattern

It was once widely believed (and some researchers still hold the belief) that the basic element in all animal behavior is the so-called reflex, that is, a motor or secretory response to a stimulus received from the environment. A regrettable consequence of this reflex theory was that it led solely to experiments designed to confirm it. The central nervous system was repeatedly exposed to new stimuli, and was hardly ever left alone long enough to show that it also could act spontaneously, in the absence of external stimuli. The basic behavior element in the neural processes of animals is never a pure response; it is a simultaneous combination of action and response.

A familiar example is provided by the stimulatory cells of the heart. If left to its own devices, the heart's atrioventricular ganglion fires rhythmically at regular intervals and can by itself make the heart beat, although at a considerably slower rate than the normal heartbeat. What makes the normal heartbeat faster is the "boss" of the atrioventricular ganglion, the sinus ganglion, which has a slightly faster rhythm. With every heartbeat, the sinus ganglion delivers an impulse to the atrioventricular ganglion just before it fires itself. When this exchange of stimuli between the sinus ganglion and the atrioventricular ganglion is suppressed, the latter pauses briefly and then continues beating in its own, slower rhythm.

The same principle applies to virtually all basic behavior patterns. They will be performed spontaneously after an extensive resting period, but these spontaneous patterns will also respond to supplementary stimuli. Even the heart's primary stimulatory centers are influenced by the nervus accelerans cordis.

As a general rule—although there are exceptions—any

given instinctive motor pattern has a specific releasing situation, what we call the *innate releasing mechanism*. If this situation fails to occur over an extended period of time, the instinctive motor pattern takes matters into its own hands, so to speak, through a lowering of the stimulus threshold, which can go so far that the motor pattern is performed in the absence of any observable stimulus. We call this a *vacuum activity*.

In fact, if an instinctive motor pattern has been dammed up for some time, the animal becomes restless. At its simplest, the restlessness takes the form of undirected searching behavior, but in many cases it leads to a directed striving for the releasing stimulus context and generates an *appetitive behavior*. The internal spontaneous drive and the external releasing stimulus are thus *summated*. What this means is that a motor pattern, whether as the result of a weak internal predisposition combined with a strong external stimulus or as the result of a strong internal predisposition combined with a weak external stimulus, will be performed at exactly the same intensity.

Because of its basic spontaneity, every instinctive motor pattern contributes to the motivation of an animal's behavior. Some of the motor patterns are highly specialized and are linked together in long chains, and others are simple, limited, reflexlike patterns. But, as I have already said, every instinctive motor pattern has "a seat and a vote in the parliament of the instincts," even the smallest elements resulting from ritualization.

This particularly applies to the multiple-purpose patterns—what I used to call the "behavioral tools"; such simple, relatively unspecialized motor sequences as walking, running, flying, gnawing, and biting, which are almost always employed in the service of special appetites. But we must remember that these frequently used minor behavioral components can generate their own action-specific appetites. A captured mouse will continue to gnaw steadily even when plenty of food is available, and a wolf in a tiny enclosure will pitifully continue to discharge its locomotor drive. The high degree of spontaneity of multiple-purpose patterns frequently leads to vacuum activities and sometimes to pathological side effects.

Because the ethologist's concern is the recognition of the motivational state of an animal subject at any given time, familiarity with the ethogram of the species is essential. Successful ethological analysis depends on a knowledge of individual motivational conditions, the varied repertoire of behavior patterns, and the competition between them.

The overall range of behavioral possibilities available to an animal constitutes the system by which it adapts to its environment. For this reason, it is impossible to treat ethology and ecology independently. In his classic *Behavior of Lower Animals*, H. S. Jennings demonstrates the systemic nature of the set of innate behavior patterns of an animal species. In the lowest organisms, such as paramecia, amoebae, or flagellates, which were the primary subjects of Jennings's research, it is obvious that an individual has only a small number of behavior patterns available. Each pattern is immediately recognizable as the function of a specific mechanism that consistently generates it, with at most some variation in intensity. The survival value of each of the mechanisms can also be easily determined. In these organisms, however, such motor patterns are too limited in number to permit any conclusions regarding their phylogenetic origins.

Characteristics of the Instinctive Motor Pattern

C. O. Whitman and O. Heinroth are rightfully regarded as the pioneers of comparative ethology. They were the first to recognize that motor patterns can be as reliable as dental formulae and feather patterns in describing the fixed characteristics of both small and large taxonomic groups of animals. (Interestingly, however, neither of them seemed aware of the physiological implications of instinctive motor patterns.) Concerned almost exclusively with instinctive motor patterns, they focused on their general properties, most notably their essential spontaneity. As far as I know, A. F. J. Portielje, the director of the Zoological Garden in Amsterdam for many years, deserves the credit for clearly identifying this property.

In 1932, I wrote a paper entitled "Observations on the Recognition of Species-Specific Innate Behavior Patterns of Birds." At the time, I knew nothing about Erich von Holst's findings regarding the endogenous generation of stimuli and central coordination, and I believed that species-specific innate behavior patterns were the product of reflex chains. Nevertheless, I had grasped the fact that instinctive motor patterns exhibit a peculiar spontaneity and that they can be performed at very different levels of intensity. I had also correctly described the above-mentioned vacuum activity, the performance of an instinctive motor pattern in the absence of releasing stimuli. Seeing a hand-reared starling in an empty room performing the complete behavioral sequence of hunting, capturing, killing, and eating flying insects, although no such prey were present, taught me the significance of this phenomenon.

The stimuli that are spontaneously generated and coordinated by the central nervous system have been aptly described by Erich von Holst as *impulse melodies*. For the recognition of an impulse melody, it is not at all necessary for it to be played at full intensity, or "loudly." There is every conceivable intermediate stage, from the lowest intensity, at which the individual sounds can be heard separately, to the complete performance of the motor pattern at full intensity. Whatever the intensity, it is always possible to recognize the characteristic configuration that we call the melody, and if we hear nothing but incomplete fragments, Gestalt perception permits us to hear the melody.

The human capacity for Gestalt perception depends on an innate physiological mechanism that permits the recognition of a chain or some other configuration of stimuli. It is a great step forward in our knowledge of the world when we can suddenly say: "This is not a coincidence." This is in line with Jakob von Uexküll's simple definition: "An object is anything that moves as a whole." David Hume's well-known axiom, which appears to exclude induction as a source of knowledge, is based on the erroneous assumption that the external world is not structured. If we can repeatedly identify a highly complex, fixed sequence in a series of external stimuli, we have no choice but to assume that the stimuli stem from a physical entity whose unchanging internal structure is responsible for the regular sequence of the information we receive.

A melody cannot be coincidental. There must always be something or someone who plays the melody. As the classical Gestalt psychologists Christian von Ehrenfels and Max Wertheimer recognized, a melody is *transferrable*. In other words, recognition of it is independent of the sound frequency, the sound quality, and the volume, and is possible even when the performance is fragmentary. The importance of this for the physiologist is that the impulse sequence of an instinctive motor pattern can also be recognized under similarly imperfect and variable conditions. The reader, noting the confidence with which we identify instinctive motor patterns, may doubt the reliability of such Gestalt recognition. But anyone who intends to take this book seriously must accept our word that we can almost infallibly recognize the frigate posture, the bent-neck posture, roll cackling, and all other such patterns.

This is not to say that we believe we know every instinctive motor pattern that a greylag goose may perform. We have had several experiences when a regular series of motor acts has suddenly sprung to our attention, although the sequence has been performed in front of us thousands of times without our recognizing it as such. Even so, the discovery of a new, highly differentiated instinctive motor pattern in the greylag goose would be no less of a surprise to us today than the recognition of an unknown butterfly species in central Europe would be to an entomologist.

The Receptor Side

Because of the practical problem of observability, an ethogram is concerned more with the motor aspects of instinctive motor patterns than with the complementary receptor or afferent aspects. The innate releasing mechanisms, which indicate to the organism when and with what intensity a given motor pattern shall be performed, can be investigated only to a limited extent through observation under natural conditions, the major source of the information presented here. Of course, releasing mechanisms can and must be studied experimentally. Niko Tinbergen has done this kind of work in his detailed investigations of the functioning, and especially the limitations, of such mechanisms. The fact that we can say so little in this book about the innate releasing mechanisms themselves should not be taken as an indication that no research has been done on them.

The Ethogram and Ecology

The word *ecology* can be defined as the study of the interactions between an organism and the environment it inhabits. (The Greek *oikos* means "house.") We cannot understand this interaction without a knowledge of the animal's body, its locomotor apparatus, its integument (the pelage or plumage), its teeth, its claws, its muscles, and so on. Similarly, the functions of the various organs of its body cannot be analyzed satisfactorily without a knowledge of the ethogram of the animal species, that is, of the set of motor patterns that the species can perform. A textbook in physiology begins with a description of the anatomical structures whose function the textbook will explain, and a textbook in anatomy begins with a description of the bones and joints. In research, it is generally a good strategy to begin with the features that most often play a causal role in the assessment of the system and the functional interrelatedness of its parts.

The available set of instinctive motor patterns and innate releasing mechanisms can be said to constitute the behavioral skeleton of an animal species. The relationships between the parts can change, just as the position of bones can change through their movement at the joints. But the changes are always confined within certain limits. A knowledge of the limits—a recognition of what an animal can and cannot do— is a precondition for understanding its interactions with the elements of its *oikos*. The description of this system is an ethogram, and the accurate compilation of an ethogram is the essential first step in the behavioral analysis of any higher animal species. It seems premature, however, to maintain that a genuinely complete ethogram has been compiled for any vertebrate species.

Compiling an Ethogram

Sources of Error

The spontaneity of instinctive motor patterns makes it necessary to take into account the action-specific potential of all motor patterns when attempting to determine motivation. As the studies of H. S. Jennings have shown, in lower organisms the small number of motor possibilities, and in particular the rarity of superimposition and mixture, facilitates analysis and makes the interpretation of motivation relatively easy. In higher vertebrates, our understanding of the motivational state at any given point is complicated by two factors. For one thing, genetically programmed instinctive motor patterns can be superimposed on each other and therefore can occur in mixed forms. In addition, the process we call ritualization complicates the analysis by developing new motivations, as the following section will show.

Learning Processes

When Pavlov's dog learned to salivate in response to the ringing of a bell, it simply meant that an unconditioned stimulus (the taste of meat) had been replaced by a conditioned stimulus (the ringing of a bell). But when a circus elephant blows a trumpet in response to a given signal, the elephant is performing a motor pattern that it would not be able to perform without prior learning. Not only primates but lower mammals can master quite a range of learned motor skills. Experimentation with various motor patterns and the learning of the skills that lead to a certain goal are known as *operant conditioning* (instrumental learning). A survey of the literature on learning plainly shows that many psychologists equate learning with operant conditioning. That is, they regard learning as no more than a process in which a motor sequence is acquired, and this process is regarded as the typical and most common form of learning.

It is important to note that the learning of new motor skills does not appear to occur with greylag geese. Through conditioning, the geese learn particular pathways, which together make up their "continental map." They learn to distinguish the physical appearance of dozens, if not hundreds, of their conspecifics, and they learn to recognize as many different plant species and their taste qualities. They learn that it is not possible to land on thin ice. In short, they learn an enormous amount, but they do not learn new motor skills. Martina never learned to slightly increase the length of her pace when walking downstairs. The only motor pattern that I have never seen in another greylag goose—and possibly it arose through instrumental learning—is the case of the goose that flew through my attic window, which was narrower than its wingspan (page 21).

Voluntary movement, the basic element through which higher vertebrates produce motor skills, is apparently not present in birds. In a series of experiments that unfortunately was never published, W. Beckwith tried to train mallards to perform voluntary head shaking. It is easy to see when a duck is about to shake its head, because it slowly raises the head feathers. At this precise point, Beckwith would throw the duck a piece of bread in order to reward the intention movement signaling head shaking. He did succeed in eliciting a few strange, uncoordinated neck movements from the ducks, but they performed no coordinated movements. What the birds managed to produce was apparently the closest they could come to developing new motor skill. Unfortunately, I learned of this very interesting result only through conversation with Beckwith.

Under the heading of exploratory behavior, I will mention only nibbling. This is the process through which the goose finds out which of the available behavior patterns—plucking, wiping, and so on—is the best for a particular foodstuff. But the behavior patterns themselves are not learned; they are definitely programmed. Learning merely determines the releasing stimulus context; it is not involved in the actual formation of the elicited behavior. Even so, there is a great deal that can be learned. As I have noted, a goose can learn an enormous number of individual characteristics of conspecifics and can remember their relative social positions over long periods of time. But the motor patterns that we see performed by those conspecifics according to their position in the rank order consist exclusively of programmed motor displays.

Rare Motor Patterns with Simple Motivation

Ethologists originally believed that the various motivational states associated with instinctive behavior patterns, which often are mutually incompatible, must of necessity occur in isolation. My mentor Julian Huxley wrote that a human being and an animal each can be thought of as a ship commanded by several captains. On the human vessel, all the captains remain on the bridge and give their orders simultaneously. Sometimes they reach a better solution in combination than when acting alone, but sometimes the conflict of opinions makes it impossible to to steer the ship properly. By contrast, on the animal vessel the captains have reached a gentleman's agreement whereby all the others quietly leave the bridge whenever a new captain turns up for duty. This analogy is appropriate in cases where the mechanism of *maximum value precedence* operates, permitting the strongest of several conflicting motivational states to have sway. The principle of maximum value precedence often applies to major categories of behavior, such as the choice between feeding and vigilance. We know of only one situation in which the conflict in the double feedback is sustained, overtaxing or suppressing maximum value precedence. This is the relatively common conflict between the motivation to fight and the motivation to flee.

At a lower level of integration, the mechanisms of different instinctive motor patterns are often in direct conflict. The conflict can be expressed to such an extent that entire organs work against each other. We find an example in the cichlid fish *Etroplus maculatus*, which, when in conflict between flight and attack, keeps its head directed toward its opponent and simultaneously tries to swim forward with the tail fin and backward with the pectoral fins. This is an exceptional case, however, since the contest between two impulse melodies usually takes place within the central nervous system. As Erich von Holst has shown, relative coordination of the various

Figure 6: *Nikolaas Tinbergen (b. 1907).*

impulse melodies generates swimming patterns that function reasonably well. We can see from this how absolute coordination might have arisen in the course of evolution, leading to the walking gait, the trot, and the gallop observed in the domestic dog.

Motivational Analysis Following Tinbergen

In higher vertebrates, the superimposition of internally motivated behavior patterns takes place in a different manner. Many years ago, Nikolaas Tinbergen (Figure 6) developed a method of identifying the independent drives underlying the performance of any behavior pattern with a mixed motivation. The first step is for the observer's Gestalt perception to come into play and produce an immediate description of any recognized motor patterns, for example in the approach or retreat of an aroused gander. This preliminary analysis of the motor patterns is followed by an analysis of their context. For instance, it may be observed that the gander's opponent sometimes adopts an aggressive stance and sometimes gives

93

Figure 7: *Threat behavior, with escape motivation increasing from (a) to (d).*

way to the attack. The third step is to record all behavior that follows the mixed motor display.

Accordingly, if a gander adopts the posture shown in Figure 7(c) we can immediately see that the posture is halfway between the postures shown in (b) and (d). We therefore assume that the bird is simultaneously motivated to attack and to retreat. If that interpretation is correct, the posture must fit the context of the moment; in other words, the behavior of the opponent facing the gander should be appropriately matched. Third, the behavior of the bird should lead as often to attack as to flight. It has been my experience that students are disappointed by the banality of this demonstration, believing they have learned nothing that was not already obvious. The threat behavior of the greylag goose is a good example of this obviousness.

In some behavior patterns, however, three or four different motivational states are present at the same time. Also, the smallest ritualized movement must be regarded as an autonomous instinctive motor pattern mixed in with the observed behavior, and this complicates the motivational analysis. Nevertheless, as in almost all attempts to analyze biological processes, it is a legitimate strategy to start by simplifying matters. Tinbergen and his coworkers followed this strategy when they began their studies of various gull species. They set out by recognizing only three possible kinds of motivation for an observed behavior, sexual, aggressive, and submissive. With the mating behavior of gulls, which is extremely simple in motivational terms, such a procedure is in fact virtually free of error. When the investigators of this group of birds

94

ruled out the possibility of a social instinct, they were basically right. The ritualized behavior patterns of gulls are also simple and can be clearly traced back to the three basic motivational sources. In these gulls, ritual plays little part in the "great parliament of the instincts."

Here, in our motivational analysis of greylag geese, we confront certain difficulties that will require our attention later, in the discussion of the hierarchy within the group. For example, any picture of forward stretching of the neck, as shown in Figure 7(a) or in the section on the motor patterns performed during aggression between rivals (see Figure 74), is ambiguous, because the slight difference in orientation that distinguishes attack from cackling (social greeting) cannot be seen in profile. In Figures 87 to 89, however, it is obvious that the recipient of the neck-stretching display is completely relaxed and that the goose performing the display is greeting, not attacking.

Ritualization

With many vertebrates—bony fish, birds, mammals—it is possible to recognize a phylogenetic process that modifies the function of instinctive motor patterns and may even generate new ones. This process and its significance were discovered by Julian Huxley as early as 1914. Functional modification usually occurs when an instinctive motor pattern that originally interacted directly with the environment is converted to a signal. An unprejudiced observer might characterize such a display derived from an original instinctive motor pattern as a ceremony, whose function is easily understood in anthropomorphic terms. When a great crested grebe fetches a bundle of nest material from the depths of the lake and presents it to his mate, anyone can read the message: "Let us go and build a nest together" (Figure 8).

Every system of communication requires transmitters and receivers. An organism's transmitted signal must be matched

Figure 8: *Great crested grebes symbolically offering nesting material to each other.*

to a complementary, phylogenetically programmed receptor mechanism that will make a species-specific response to the signal, a response that promotes survival. The evolution of such a system usually begins on the receptor side, starting from a particular motor pattern that exerts an "infectious" effect. In comparative psychology, this is often referred to as social induction—a vacuous term that only gives the impression of being an explanation. The comparative physiologist would say that a particular receptor mechanism has been preprogrammed to fit the external stimulus. Wolfgang Wickler has called it *semantization* (from the Greek *sema*, meaning "sign"). Semantization results in selection pressure exerted on all components of the behavior pattern involved, both reinforcing and refining the signal. The selection pressure, in turn, leads to the formal exaggeration of many motor patterns. The classic example of this process of ritualization is the very familiar food-summoning call of the domestic chicken. As with many other birds, feeding in chickens has an infectious effect brought on by the tapping of beaks on the ground—

something every farmer knows. J. Effertz kept young chicks on a surface that produced a strong echo when pecked, and this significantly increased their feeding motivation. In domestic chickens, both the leading hen and the cock reinforce pecking with a vocal accompaniment that Wilhelm Busch has translated as *tack, tack, tack*—"Here we come."

A typical example of ritualization in greylag geese is the bent-neck display. In its complete form, the pattern constitutes an invitation from the gander to the goose to accompany him step for step in the daily search for food. In the nonritualized version, the gander alternates between the sneaky posture, which he assumes consistently while walking along, and the pattern of looking downward when searching for food. As the result of a process that we do not yet understand, these two motor patterns have become fused, and the phenotype of the combined motor pattern has become fixed as an exact copy that is genetically programmed. Geneticists recognize the process of phenocopy formation, through which the chance combination of internal and external stimuli can produce a movement that is the same as one fixed in the genotype. In other words, if we observe a goose walking along in the sneaky posture, glancing down once or twice, and holding its beak almost vertically, it has no particular significance. But if we repeatedly see the combination of sneaky posture with downward looking performed for seconds at a time, and if it is consistently associated with the partner walking in parallel, then we must conclude that a new behavior pattern is in operation. Such fixed combinations of independent motor elements are regularly found in ritualized behavior patterns.

While the bent-neck posture as a ritualized pattern is clearly distinct from its nonritualized components, there are other behavior patterns that show every conceivable intermediate stage between the nonritualized and the ritualized form. Neck dipping by the gander, which serves as a prelude to mating in its ritualized form, has undergone only mild

97

modification in the process of ritualization. If we watch a gander upending to feed on the roots of reeds, we correctly identify the neck dipping as a basic behavior pattern. If, in the process of dipping his head in the water, the gander takes a sideways glance at a nearby goose, we still would not regard it as a ritualized pattern. Accompanying indications of interest in a female can increase, however, and we have no way of calculating, in the mixture we observe, the relative roles played by the original upending motivation and the courtship motivation.

Imprinting

Imprinting is a developmental process by which behavior becomes attached to a particular object. It is distinguished in a number of ways from other types of developmental change in behavior. In the first place, no reinforcement is necessary; mere passive exposure of the individual to a particular stimulus situation is enough to ensure attachment. A second characteristic of imprinting is its irreversibility, or at least its strong resistance to a reversal of the process. A third feature of imprinting is its limitation to a specific phase of development, which often lasts only a few hours. We have already encountered some examples of imprinting in the fixation of the following behavior of goslings. We will not be concerned here with imprinting's effect on the maturation of behavior, which can lead to several other imprinting processes.

A peculiar and hard-to-explain property of imprinting is that it is always related to the species and not to the object that transmits the imprinting stimuli. Friedrich Schutz imprinted male mallards on shelducks and on members of other duck species by keeping the young birds for a number of weeks with the duck on which they were to be imprinted. The experimental birds were then set free on Lake Ess among hundreds of ducks of different species. The following spring, the mallard drakes invariably selected a partner of the species on which they were imprinted, but they never selected the individual with which they had grown up.

The most bizarre example of the imprinted individual's generalization of the imprinting species is provided by the first jackdaw I hand-reared. When the bird reached sexual maturity two years later, it fell in love with a sweet, dark-haired young girl from the next village. It is a complete mystery to me how the jackdaw was able to recognize that two such different people belonged to the same species.

The twin nature of the developmental process—the early, irreversible attachment to the species of the caretaker and the reversible recognition of that person as the imprinting individual—has great importance for our goose-training procedures. A goose whose following response has been fixated on one human can easily be taken over by another, especially if its responses are gradually shifted by having the substitute lead the goose along with the original caretaker for a few days. By contrast, a gosling that has followed a mother of its own species for only a few minutes cannot normally be induced to follow a human. When one is aware of the great power exerted by learning and habit in other contexts, the final and irreversible nature of imprinting is a constant source of surprise.

The developmental stage at which the imprinting phase occurs does not seem to bear any relationship to the time at which the resulting behavior pattern will be performed for the first time. Imprinting can take place minutes, months, even years before the relevant behavior is elicited. For example, the phase during which the sexual responses of the jackdaw are imprinted occurs some time before the phase in which the following response is fixated on a parent or a human caretaker.

The stimuli that an object must transmit in order to serve as a focus for imprinting obviously vary greatly among different animals and different behavioral systems. As Eckhard Hess has demonstrated experimentally, imprinting in the mallard requires that the object be approximately mallard size and move away at a particular speed when the duckling is in the psychological state of distress-calling. Peter Klopfer has observed with the wood duck (*Aix sponsa*) that imprinting takes place even before the ducklings leave the brood cavity. It is brought on by the repeated exchange of distress calls from the ducklings and contact calls from the mother. A dialogue of this kind probably plays a part in the greylag goose as well.

When I was setting up the new installations in Seewiesen, I constructed a device that was intended to direct the imprinting of young greylag geese to a substitute object in the best way possible—without human intervention. I rigged an artificial hen, equipped with a loudspeaker and attached to a long arm, to move in circles inside a large cage. The goslings exposed to this dummy mother did indeed learn to use the artificial hen as a source of warmth, but they showed no following response, whereas the ducklings in Hess's experiments had followed a rotating dummy without any problem. When I examined my own behavior toward the goslings, I realized to my shame that I had overlooked one vital point— a human being responds quite involuntarily to the distress calls of goslings. But one of the key factors in their imprinting is that the vocalizations of the imprinting object must be produced in *response* to their distress calls.

If the imprinting process is eliminated by rearing animals in virtual isolation (as in the Kaspar Hauser experiments), these geese will avoid conspecifics and refuse to have anything to do with each other. When two such geese are placed in a cage together, they fall into the habit of sitting in opposite corners, as far away from each other as possible. Their responses to conspecifics are remarkably unpredictable. The overall picture of such disturbed behavior resembles that of the autistic human. During a brief discussion, Fritz Riemann, a psychiatrist practicing in Munich, once asked me: "Are your Kaspar Hauser geese tactless?" Helga Mablona-Fischer and I looked at each other in amazement, for that exactly describes the behavior of these experimental greylag geese. They mis-understand displays; they will attempt, for example, to court powerful ganders, who immediately respond with violent aggressive displays. Initially, we took such misunderstanding of the displays of conspecifics as an indication that, while the motor patterns of the goose's ethogram are innate, their meaning must be learned. Later experience showed that that interpretation was wrong. Greylag geese understand the dis-plays and vocalizations of their conspecifics just as innately as they perform the behavior patterns. That is, they do not depend on any learning processes for behavioral recognition. The fact that the Kaspar Hauser animals shut out all stimuli coming from conspecifics is actually the result of a funda-mental disturbance of their behavior.

I was able to rule out our original interpretation through a simple but somewhat demanding experiment. Immediately after hatching, three artificially incubated goslings were given to three different people who were extremely good with animals. In raising the goslings, they were to be careful to develop social contact with them but were to keep them strictly isolated from other geese. The three geese were set free in Seewiesen when they were approximately one year old. They showed no sign of the "tactless" behavior of the Kaspar Hauser animals and immediately joined the other geese. Contrary to our expectations, two of them (Wipa and Inga) paired to-gether. This probably came about because, in the attempt to maintain their contacts with humans, they frequently turned up at the children's playground and there became acquainted with each other. Wipa, a gander that had grown up in the parking area of the Mittelstaedt Institute in Seewiesen, exhib-ited one minor imprinting abnormality. Although he showed perfectly normal sexual responses, his aggressive behavior was clearly fixated on the bumpers of Volkswagen cars—a fixation that eventually led to his death.

We are extremely fortunate to have chosen greylag geese as our experimental subjects, because the natural drives of those reared by humans are influenced only to the extent that they will accept humans as substitute parents and later as social partners. Some closely related species, such as the bean goose (*Anser fabalis*) and especially the barnacle goose (*Branta leucopsis*), have a strong tendency toward perverse behavioral responses when they are reared by humans. Hand-reared barnacle geese often show very different behavior than those reared by their biological parents.

Ethogram 1

The complete ethogram of an animal species includes a great variety of behavior patterns of differing degrees of complexity. The differences occur on both the afferent and the efferent sides of the behavior pattern; that is, the stimulus that elicits a motor pattern and the motor pattern itself can vary. The degrees of complexity of the releasing mechanism and of the motor pattern it releases are completely independent of each other. There are behaviors consisting of a single instinctive motor pattern and involving only one coordinated movement that are elicited by many different combinations of stimuli. Some examples are distress-calling and the eyelid-closing reflex. The latter, a very simple behavioral response, occurs every time a greylag goose sees an object that might come in contact with the surface of its eye, no matter what the shape of the object.

On the other hand, there are behavior patterns in which many of the motor patterns preprogrammed in the central nervous system combine to form a functional unit. Eliciting such a pattern can be simple, or it can be extremely complex. It is quite impossible to draw a sharp line between simple behavior patterns and those containing many components. Rather, there is a continuum from the simplest to the most highly integrated, both in the releasing mechanism and in the observable behavior. The individual components of an ethogram are therefore not always listed one beside the other; sometimes a hierarchical arrangement is better. Also, the descriptions for the different components may correspond to different physiological levels. Erich von Holst conceived the idea of using a special terminology to reflect these levels. Different physiological levels, in turn, are related to different functional goals. The courtship of a female or the leading of offspring requires behavior patterns of a higher order than behavior that is meant only to whisk away a pesky fly.

Hatching

The earliest behavior patterns we can observe in a young greylag goose are some that are performed while it is still in the egg. Just before it hatches, the gosling communicates with its parents—for the father now stays very close to the nest. It seems likely that the father has been alerted by vocalizations from the young in the eggs, although it is possible that the mother communicated with him in some way.

During the incubation period of twenty-eight to thirty days, the growing embryo draws nourishment primarily from the albumen of the egg white. The yolk is not used until just before hatching and for a short period afterward. The air chamber in the egg gradually becomes larger, and the embryo develops to fill the space between this chamber, at the blunt end of the egg, and the still untouched yolk. The embryo's body takes a specific position within the egg. The head is bent sharply forward at the neck, so that the nape of the neck points toward the blunt end of the egg. The head is held in the armpit under the right wing, bringing the beak into the plane of the back and shoulder. This leaves the upper surface of the beak, the egg tooth, and the top of the head pressed against the shell on the right (Figure 9). We have never investigated whether a mirror image of the hatching posture can occur. The tip of the beak, with the egg tooth, touches the shell of the egg very close to the point at which the egg membrane is attached. The nape of the neck and the top of the head are next to the air chamber. When the egg tooth

Figure 9: *The position of the gosling* *in the egg at the beginning of the hatch-* *ing process. (This gosling died shortly* *before hatching.)*

Figure 10: *Pressure from the egg tooth* *breaks the shell of the egg.*

breaks through the membrane, the beak and the nostrils are immediately thrust into the air chamber. Breathing with the lungs begins, and the first vocalizations are then produced. The unhatched gosling can already utter the distress call, the contact call, and the sleepy call. The distress call is elicited when the egg cools or if obstacles are encountered during the hatching process—for example, if the gosling becomes trapped by a dried piece of the egg membrane. When a call of the appropriate frequency is given in reply, the gosling first responds with a single-syllable contact call, which later becomes multisyllabled. When a cooling egg is rewarmed, the trilling sound of the sleepy call can be heard. The goslings can produce all three vocalizations as soon as the nostrils have penetrated the air chamber, before the shell is broken.

An observer might conclude from the sounds that the gosling is hatched and may attempt, in his concern, to remove the eggshell. Under no circumstances should this be done, since the yolk still has a diameter of 2.5 to 3 centimeters and is exposed beneath the belly. Only after the gosling has used its lungs to breathe and to produce peeping calls for some time does it begin to break through the eggshell (Figure 10),

Figure 11: *The rotation of the gosling results in a circular break around the eggshell.*

by exerting pressure from the egg tooth with its neck muscles. The egg tooth, a genuine tooth covered with enamel, is found on the tip of the snout in most reptiles and almost all birds. The common expression that the shell is "pecked" is misleading, as the bird has no room to jab with its beak. Instead, it presses outward with the egg tooth, which it holds firmly against the inside of the shell. The first pieces to break in the shell always point outward. The movement necessary to press the tip of the beak outward also rotates the embryo around its longitudinal axis. As a result, the tip of the beak makes a circle of breakages around the shell (Figure 11). When, to use Heinroth's expression, the "tropic of Cancer" has been broken out, the hatchling stretches its neck and back and lifts away the "northern temperate zone." As it does so, the head falls free over the edge of the remaining piece of the shell. But the yolk sac has not completely collapsed yet, nor has the umbilical cord been sealed off.

Up to this point, the extensor muscles of the neck, shoulders, and back have done the main work involved in hatching. Now a critical stage has been reached, because the umbilical ring must be moved forward over the egg membrane. The umbilical cord will have closed by now, and the umbilical ring, consisting of a broad wreath of gelatinous embryonic tissue, will rapidly dry and shrink once it is not pressing against the membrane. This is a stage at which

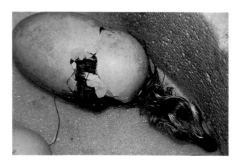

Figure 12: *The gosling pushes away the blunt end of the egg by stretching its body. As soon as the head is free, the balancing organ begins to function, so that the crown of the head is oriented upward.*

disruption of the hatching process is likely to occur. A stretching movement with both legs now pushes the young bird forward and farther out of the shell. The sensitive umbilical ring touches the nest material, and within a few minutes it shrinks completely.

The gosling is unable to lift its head, but it can hold the head with the crown balanced and pointing upward (Figure 12). If something is said to the gosling in an appropriate voice, it will try to lift its head by stretching its neck. In other words, the gosling is already trying to perform a greeting. As soon as the head can be carried even slightly, we may see the shelter-seeking response. That is, a gosling lying exposed in the open will try to move into the shade by pushing with both legs.

The newly hatched gosling looks wetter than it actually is. That is because every down feather is enclosed in a thin horny sheath, which becomes brittle as it dries and falls away when the gosling begins to move around. The down feathers at once unfold, and friction between the mother and the gosling gives them a charge of static electricity, which makes the tips of the down feathers repel each other and stand away from the body a regular distance apart (Figure 13). This accounts for the fluffy appearance of the gosling, and it also ensures

Figure 13: *The feather sheaths break away as they dry, and the down feathers appear to become several times larger. The egg tooth, which does not fall off for several days, can be clearly seen in this photograph.*

the waterproofing of the plumage. Goslings hatched in isolation must be rubbed with a pad of cotton wool to generate the static electricity.

Because geese and other birds interrupt the hatching process at night in order to rest, it is not possible to state the precise age at which a gosling can stand on both legs and walk. We usually see it during the first morning after the day of hatching. The gosling, standing in front of its mother on the edge of the nest, looks around in all directions. If the mother goose remains silent and unmoving, the gosling utters distress calls. The mother responds with the contact call, and shortly this exchange ensures the process of imprinting that fixes the species but not the individual recognition of the parent. At this age, greylag goslings will indiscriminately join up with other goslings or attach themselves to strange parents.

They can also be induced to greet an organism that does not belong to their species.

After hatching, the goose removes the remains of the shell. If this process happens to go wrong, the blunt egg cap can fall on an egg that is not so far advanced and suffocate the hatching gosling inside.

Locomotion

Running and Walking

We do not know whether the ancestors of birds walked by alternating the movements of their legs or whether they hopped along. It is quite possible that tree-living reptiles performed two-legged hopping from branch to branch before the lineage leading to birds developed. I am inclined to believe that alternating walking was the primitive pattern of movement in the ancestry of birds. The only conflicting evidence for this hypothesis is that studies of the development of locomotion in birds have not revealed a single species in which the adults hop on two legs but their developing offspring walk with alternating leg movements. The evidence from ontogeny is ambiguous, however. For example, we know of some passerine birds, such as ravens and larks, that show two-legged hopping for a short while after leaving the nest but later walk with alternating leg movements. These birds do not show a genuine trot.

In geese, the usual form of locomotion is slow walking with alternating movements of the legs and with both feet never off the ground at the same time. When greatly aroused, as during attack or flight, they switch to running (Figure 14). In anatids, gallinaceous birds, and many other groups (especially the ratites), running is a genuine trot. The transition from the walking gait to the trot is as clear as it is in long-

Figure 14: *Walking and trotting.*

legged mammals. By contrast, the running of songbirds is not really a trot but, rather, a much faster version of the walking gait. What apparently plays a large part is the very long, barely curved claws on the rear toes that are found in the fastest runners, such as pipits, larks, and wagtails. I define genuine trotting as a movement in which the thrusting foot leaves the ground before the advancing foot makes contact, so that the bird completely leaves the ground with each pace.

All the anatids with which I am familiar, even those that are most fully adapted for swimming, can perform some kind of a trot, although their walking may be only a slow waddle. Long-legged species that are specialized for grazing exhibit a well-developed trot. The Cape barren goose (*Cereopsis novae-hollandiae*) is especially well equipped for this form of loco-motion. Greylag geese also have a rapid trot, and it is worth mentioning that the to-and-fro waddling movement of the long axis of the body that is so evident in normal walking completely disappears in the trot. The trotting speed is so fast that a running man has difficulty keeping up.

The fastest running speeds are usually seen when molting geese are heading for the water or toward cover. While the primaries and secondaries are still in their vascular sheaths, geese show an extremely strong inhibition against using their wings at all. The vascular sheath is apparently quite vulnerable, and any damage to it can lead to permanent deformation of the feather. When a timid greylag goose that is molting its pinion feathers is chased, it may go leaping down a slope for several meters without opening its wings.

The behavior patterns involved in rapid escape on foot while protecting the wings are retained unchanged even by geese that have been pinioned and have been unable to fly for many years. It is almost impossible to induce the maximum running speed in geese outside the molt because then the motor patterns of takeoff are elicited before the top speed is reached, and this abruptly interrupts the running sequence. That is why it is easier to catch amputated geese outside the molt than when they are molting. Suppression of the takeoff response through learning, which comes only after a lengthy period without the capacity to fly (as in zoos), is far more effective in geese and swans than in smaller, "less intelligent" anatids. For example, amputated teal never learn that they cannot fly.

Geese can also perform a two-legged leap. If a solid obstacle up to about shoulder height is blocking the way, the goose takes heed of it from several paces away and fixates the top with its head tucked in and the neck trembling. The goose then stops in front of the obstacle, fixates the top of it with both eyes, and leaps accurately onto it with simultaneous thrusts of both legs. If the top of the object is narrow, the goose immediately jumps down on the other side. Greylag geese are perfectly able to remain standing on large rocks and branches, but on thin branches they can only perch precariously with jerky balancing movements. By contrast, wood ducks and their relatives cope very well with such perches, and in Slimbridge I have often seen magpie geese (*Anseranas semipalmata*) resting on telephone cables.

The two-legged leap is rarely seen with very narrow obstacles that are known to be fixed, such as an iron fence surrounding a lawn. But we often observe another form of spatial orientation, which is interesting because the receptor side apparently yields more information than the animal is able to exploit in its motor response. While the goose is still several meters away, we can recognize its intention to surmount the obstacle. Well before it reaches the obstacle, the goose fixates it, with its head tucked in and the neck trembling, so that it knows exactly where the obstacle is. But a limited mastery over its motor apparatus forces it to perform a peculiar pattern of movement. The approaching goose begins to raise its feet higher and higher, sometimes raising them higher than necessary even before the obstacle is reached. It is only by chance that a foot falls on top of the obstacle; equally often, it falls short or goes too far. When the foot does go too far, so that the rear of the leg touches the top of the obstacle, the goose flaps its wings a little to help it hop over.

There is a second functional miscarriage that is of interest in relation to the spatial orientation of the goose and its capacity to adjust voluntary movements. Here, too, the goose's processes of spatial orientation can achieve more than can the adjustment of the motor activity that the process controls. My greylag goose Martina was well able to walk, with some effort, up our staircase, which had broad, carpeted steps. While still a half-grown gosling, she could adjust the length of her paces to the distance between the steps, but only when she was walking upstairs. She was quite unable to do this when going down. The distance between the steps was somewhat too far to fit the natural length of her stride. To solve the problem, Martina should have either lengthened her stride or offset the disparity by slipping in a small extra pace. That is exactly the response that tree-living anatids such as Muscovy ducks, wood ducks, and mandarin ducks would have shown, yet it is not shown by the greylag goose, which is in other respects much more "intelligent." When the difference between her stride and the distance between the steps was so great that she was unable to reach the next step down with her foot, Martina did not know how to proceed. She would pull back her outstretched leg after it failed to make contact with the next step, but then she would simply stretch the leg out again. Before fledging, she would go no farther once she had reached this point, but would stop and begin to utter distress calls. Later, she would fly down the last difficult stretch of the stairs, and soon she was able to avoid the stairs altogether by flying out my window to leave the house.

Swimming

Unlike many mammals, the movements the greylag goose uses in swimming are quite different from those used in walking. The difference in the motor patterns becomes especially clear when we see the way a goose switches from walking to swimming as it launches itself on the water. The goose leans forward toward the water and settles its extended chest on the surface. The greater the height between its foothold and the surface of the water, the more the neck is pulled back. The shanks are directed horizontally backward to bring the ankles approximately level with the water surface. Then, abruptly, the buoyancy of the water supports the body, sometimes with an audible plop. One amusing observation we made at a small stream bridged by a narrow plank a few millimeters below the surface helped us to analyze this motor sequence. A greylag goose showed a remarkable functional miscarriage in using this bridge: it walked onto the plank, settled down on it with the motor sequence for launching, and began to paddle with its feet on either side of the plank. Heinroth observed a similar functional miscarriage with mute swans that had to cross a pool a few millimeters deep lying on the surface of a solid layer of ice.

If the water is deep enough to carry the goose's body but too shallow to allow the legs complete freedom of movement in paddling, the bird can propel itself forward with skilled movements that are adjusted both to the bottom and to the depth of the water. Greylag geese rarely encounter fast-flowing water in their natural habitat, and it is our impression that our geese considerably improved their movements in this context through learning. During unobstructed swimming on open water, the legs move in alternation, the foot stretched so far backward that it almost breaks the surface. The movement can be interrupted at this point so that the goose glides along with one leg stretched backward and the other held against the body in an extreme forward position. In the foot that is pointing backward, the middle and inner toes are usually held together and the outer toe stretched upward, so that the web serves as a sort of vertical rudder.

The frequency of the leg movements can be increased

only so much in the goose's attempt to go faster—far below the maximum frequency seen in running. Once the maximum swimming speed has been reached, a new motor pattern appears: the goose tries to raise its body above the surface and appears to run on the water with rapid leg movements. This running on the surface is important for young goslings and even more so for ducklings. It can be kept up over fairly long distances, in relation to body length. An adult greylag goose uses its wings to assist in this kind of locomotion, especially when fleeing rapidly, as after its defeat by a conspecific.

With the exception of the mute swan and perhaps the magpie goose, all anatids can dive. In the true diving ducks (Aythyini) and certain other species that are specially adapted for diving, such as members of the genus *Oxyura*, the wing remains tucked into the wing pocket while underwater. By contrast, geese spread their wings (which are otherwise kept folded) sideways from the shoulder joint during diving. This allows the membrane stretching between the upper arm and the forearm, the propatagium, to press against the water and help drive the goose's body (especially the front of it) downward. The feet, held with the ankles raised almost level with the back, also exert a strong downward thrust.

Geese dive only when fleeing or when engaging in play diving, never when feeding. Any food that lies below the reach of the beak during upending remains untouched. This gives any vegetation that grows in a goose pond, such as *Elodea*, the appearance of being cropped down to a certain level.

Not just waterfowl but many birds possess an innate pattern of coordination of wing movements for swimming. The pattern is quite distinct from that of flying. Instead of moving around an axis that is roughly parallel to the long axis of the body, the wings move to and fro around an almost vertical axis perpendicular to the flight direction. That is to say, a swimming bird spreads its wings without raising them high above the back, stretches them well forward, and dips them into the water. The wings are then pulled back in a powerful but rather slow rowing action that drives the bird forward through the water. To my amazement, I repeatedly saw my ravens perform this motor pattern while bathing on the bank of the Danube. They would often lose their footing while bathing and be swept away from the bank. Showing no sign of alarm, they would row themselves back to the bank with the movements just described and resume their bathing as soon as they found a foothold.

This motor pattern, which appears to have a long phylogenetic history, is often observed with anatids. We can evoke it at will from domestic ducks that are unable to fly simply by chasing them. With geese, the pattern is regularly included in the varied sequence of escape movements they perform during their daily bathing activity. Heinroth used to say testily that once the bird's feathers are wet, it automatically goes on to perform all the motor patterns connected with wet feathers. The swimming pattern has been specially enhanced in the South American steamer duck (*Tachyeres*), which owes its name to the loud noise it produces in the process.

During the so-called play diving that regularly occurs around midday, the geese perform a wide variety of locomotor behavior patterns in jumbled sequences (Figures 15, 16, and 17). A goose may dive downward, frequently while sending a jet of water high in the air, resurface, throw in a short burst of running across the water, and suddenly dive again and zigzag through the water, abruptly changing direction. A goose can even burst through the surface like a leaping fish and take to the air, either flying off or plummeting back into the water a few meters away, like a tern. I have never witnessed an attack by a sea eagle on a swimming goose, but I am convinced that these motor sequences are intended as a defense against this predator.

It is a question whether the mixture of escape responses and bathing activities should be labeled as play or not. The

Figures 15, 16, and 17: *Play diving shows jumbled sequences of various locomotor behavior patterns.*

performance in rapid sequence of movements with different motivational backgrounds gives the impression of play, but the individual motor patterns that are performed look too much like fixed vacuum activities to fit this impression. I have also been unable to identify any component of explorative behavior, which is what one would expect in true play activity.

Flying

Very few people, including some acute observers, notice that when a large bird is flying freely, its body does not begin to sink even during the phase in which the wings are swung upward. Because the upward-swinging wing surface does not seem to provide any support, it would be expected that the bird would be pulled downward by the force of gravity. The phenomenon has a simple explanation: even when the wings are swung upward, the headwind exerts enough pressure to keep the bird from sinking. A slight rotation of the wing, so that its leading edge is tilted upward, is enough to produce this effect. The result is that the bird's entire flight apparatus is in a gliding condition during both upward and downward beats of the wings. I have accordingly termed this form of flight swing-gliding (Figure 18).

The diagram in Figure 19 illustrates the mechanics of swing-gliding when the bird is generating no upward thrust and is gliding downward as rapidly as if the wings were held still. Figure 20 shows what happens when the bird tries to gain height rapidly. In both figures, the path followed by the center of gravity is a dashed line, and the path followed by the wing cross-section is a continuous line. When the bird is gaining height, it raises its body parallel to the longitudinal axis without producing any marked acceleration or braking effect. The bird behaves something like the skier in Figure 21, who slides downward at a fixed angle while maintaining his height on the slope by stepping sideways.

Quite different mechanics are applied in flapping flight,

Figure 18: *Swing-gliding geese. It can be seen that in the course of this motor sequence the body hardly descends when the wings are swung upward.*

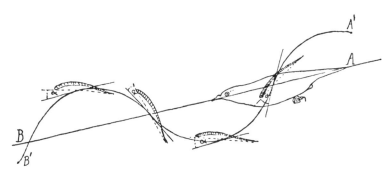

Figure 19: *Diagram of idling wing beats during swing-gliding. The wing is moved along the sine wave A'B' without performing any work. At every point, the wing is oriented at a standard angle α to the sine wave.*

when the bird flies without a headwind or any other air movement. Here, instead of being held passively against the air pressure and being driven upward because of its resistance, the wing is actively beaten against the air. The gaps between the individual wing feathers open up and are, of course, most

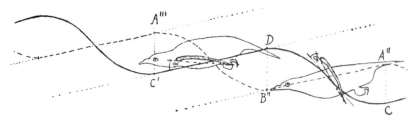

Figure 20: *Diagram of the steepest trajectory in swing-gliding. The path of the center of gravity is shown by a dashed line and that of the wing cross-section by a continuous line.*

Figure 21: *A skier taking sideways steps while sliding forward and downward. This corresponds to the way in which muscular work is converted into potential rather than kinetic energy during swing-gliding.*

pronounced near the wing tip, which moves the fastest. As a result, the greatest forward thrust is generated at the point farthest away from the body (Figure 22).

In swing-gliding, the bird is driven forward roughly in the direction of its body axis. In flapping flight, the axis of the wing beat can be modified in such a way that the forward thrust can be oriented in almost any direction. We can see from the proportions of a bird's wing the relative importance of each kind of flight in its natural behavior. Specialists in flapping flight always have short upper arms and long hands with well-developed primaries. The humerus rotates mainly around its longitudinal axis during the wing beat, and the

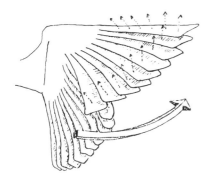

Figure 22: *In flapping flight, the upward beat of the wing is active. The small arrows show the approximate direction of the pressure operating on the upper side of every primary feather. Note the displaced position of the innermost primary feather.*

oscillating masses of the musculature and skeleton are arranged as closely as possible along the wing-beat axis.

For this reason, small birds are generally unable to perform swing-gliding. After a brief bout of flapping that provides acceleration and uplift, the wings are folded completely and the bird falls downward like an arrow until the next bout of flapping. Gliding with the wings spread apparently is not suitable because the braking effect exceeds the saving of kinetic energy while losing height. The only small birds that glide are swifts and swallows. Starlings will also glide in the unusual circumstance in which a rich supply of flying insects makes it worthwhile for them to remain in the air, as I once saw with a swarm of mayflies (Ephemeridae) above the Danube.

Birds that have a very long humerus, with the primaries away from the axis of the wing beat, are virtually unable to perform flapping flight. When there is no air movement, albatrosses must make a long run-up on foot before they can attain the headwind necessary for swing-gliding. On the other hand, their extremely low wing weight makes them able to take to the air even in a very mild breeze without flapping their wings. I have seen petrels (Procellariidae) take off in a slight air current without beating their wings but by spreading

Figure 23: *Flapping flight. The entire hand and all the individual primaries are oriented so that the forward thrust operates upward along the axis of the wing beat.*

them against the wind and driving themselves forward with a few rowing movements of their webbed feet.

Geese occupy an intermediate position with respect to their ability in the various kinds of flight. They use swing-gliding to fly long distances and flapping flight during takeoff and when braking (Figures 23, 24, and 25). Because, as has been noted, the forward thrust during flapping flight can be oriented in any direction, this type of flight is also important for landing, especially for landing on a particular spot (Figure 26). To save energy, however, the geese perform flapping flight as little as possible, and for the same reason they orient themselves against the wind during takeoff and landing. To take off, greylag geese usually run forward a few paces before they start the full flapping flight (Figure 27). If a suitable water surface is available for takeoff, they avoid some of the energy expenditure of flapping flight by running on the water

Figures 24 and 25: *Downbeat during flapping flight. The air pressure acts against the entire undersurface of the wing.*

Figure 26: *Wing flapping during point landing.*

to build up the necessary momentum (Figure 28). Some long-legged birds always take off in this manner, which has given rise to the erroneous belief that they cannot perform flapping flight.

In emergencies—for example, when frightened by a strange dog in a courtyard surrounded by houses—geese are perfectly capable of flying straight upward to the height of a three-story house, although this obviously leaves them out of breath. When a large number of geese standing close together on dry land become greatly alarmed, they all tend to use flapping flight to take to the air together. Surprisingly, collisions between them are rare. Geese that are taking off simultaneously usually move apart along a line perpendicular to the wind direction, giving every individual a clear path.

With greylag geese and their relatives—but not with snow geese—we can hear a peculiar rasping or rattling noise during

Figures 27 and 28: *In the takeoff from either land or water, the forward thrust generated by flapping flight is augmented by leg movements.*

Figure 29: *When flying over long distances, the flock takes up the familiar V formation. It does not seem to bring any aerodynamic advantage, except that each individual can avoid the tail stream of the goose in front.*

certain transitional phases from flapping flight to swing-gliding and vice versa. I heard this noise at its loudest one time when I was drifting down the Danube in a boat in thick fog toward nightfall. A flock of geese appeared above me and suddenly tried to fly steeply upward in panic. For many years I did not know the reason for their noise, until I saw the connection through a favorable combination of light and angle of view. Apparently the air current breaks away from the upper side of the wing and causes the upper wing coverts of the secondary feathers to rise vertically from the wing and begin to flutter. That is what produces the noise.

A large flock of geese flying a long distance often adopts the pattern of an acute-angle triangle, the so-called V formation (Figure 29, Plate XVI). There has been much discussion over the possible function of this formation. It is certain that the following geese do not gain any aerodynamic advantage from the work of those ahead, but it is possible that one

Plate I:
A gosling distress-calling. All the features of the call that evoke an emotional response from humans have been slightly exaggerated.

Plate II:
(1) Auingerhof, showing the institute building and the autumn and winter quarters of the geese.
(2) Lake Alm, the breeding and molting grounds.
(3) The ponds at Oberganslbach, the rearing grounds and summer grazing area.

Plate III:

(1) When swimming rapidly, the goslings follow the parents in single file.

(2) Immediately after leaving the nest, the goslings keep closer to each other than to the parents.

Plate IV:

(1) The gog *call, the commonest alarm call of the greylag goose.*

(2) In hissing, the channel between the tongue and the palate is narrowed by the forward movement of the hyoid bone. Note the bulging eye, signifying the arousal of the sympathetic nervous system.

(3) Nest-calling. The goose is staring with both eyes at the source of the disturbance.

Plate V:

(1) *A male bean goose attacks M. Martys and sparks gog calling. The infectious effect of the call leads through escalation to mobbing.*

(2) *Greylag geese, some bean geese, and a flock of domestic geese, all mobbing an otter that is visible only from the ripples it makes on the water.*

Plate VI:
(1) A nest on an island.
(2) A nest in the woods.
(3) A nest in a nesting box.
(4) A timid goose crouching when disturbed on the nest.

Plate VII:

Typical nest sites on Lake Alm:
(1) A nest shielded by sparse vegetation.
(2) An exposed nest on a tussock.

Plate VIII:
(1) A clutch of eggs in a nest lined with down feathers.
(2) A goose transferring nest material on the site that has been selected.

Plate IX:
(1) and (2) The egg-rolling pattern. The egg is balanced on the underside of the
beak as it is rolled back into the nest.
(3) Hollowing out the nest.
(4) Turning the eggs ensures uniform incubation of the clutch.

Plate X:

(1) When the goose leaves the nest for a brooding pause, she covers the eggs first with down feathers and then with the less conspicuous nest material.

(2) After returning to the nest, the goose tucks nest material under her body to ensure better heat insulation.

(3) The goslings are sheltered under the mother's wings.

Plate XI:

(1) and (2) A wing-shoulder fight. Each gander grasps his opponent by the shoulder and tries to force him to the ground so that he can strike with one of his wings, which is bent at the wrist joint.

Plate XII:
(1) and (2) In this wing-shoulder battle, the horny fighting spur on the wrist joint of the wing is clearly visible.

Plate XIII:

(1), (2), and (3) Fighting among goslings within a family to establish the rank order. The motor pattern is the same as that involved in the wing-shoulder battles of adult geese. The first picture shows how one wing is stretched backward to maintain balance.

(4) The parents hiss at their fighting offspring while fixating them with both eyes.

Plates XIV and XV:
The motor sequence of the triumph ceremony. Following an attack on land or on the water, the gander utters a rolling call while returning to his mate and ends with pressed cackling.

Plate XVI:
(1) A migrating flock in V formation.
(2) Breakup of the V formation before landing.

individual can avoid the tail stream of another by not flying directly behind it.

I myself am inclined to accept another explanation. Swimming geese never follow each other in single file, but swim just enough to the side of the goose in front as to have a clear view ahead. The pattern is barely detectable among geese walking on dry land because of the unevenness of the terrain, but it is obvious when they are crossing a large lake. In taking off, each goose moves sideways just enough so that every individual has a clear view past the tip of the outstretched wing of the bird in front.

The goose flying at the head of the V formation is not necessarily the leader of the flock. The visual field of a goose extends so far backward that it can easily see all the other individuals in the formation. But because of the bird's incredibly short response time, it has been impossible for a human observer to determine which of the geese in the formation is giving the commands to change direction. It is likewise impossible to spot the leader in a flock of plovers or teal.

Flying Downward

When geese arrive high in the sky—usually only after they have flown a considerable distance—they stop beating their wings long before they arrive over their intended landing site and start to glide with their wings fully spread. Reducing the wing surface would lead to a sharp acceleration in speed, which is undesirable just before landing, and so the birds have to use other means to lose height. One spectacular way they do this is by flying upside down (Figure 30). The goose flips over on its back so that the wing surface is directed into the headwind from below, and it dives downward at double speed. Upside-down flying is presumably a property of all flying birds with a breast keel (Carinatae), and in some breeds of pigeon it has been artificially selected to an almost pathological degree. The survival value of this motor pattern is no doubt related to the escape from aerial predators that attack

Figure 30: *In order to lose height rapidly, for example when escaping from a raptor attacking from above, a flying goose can turn over on its back.*

from above. Through the entire performance the head remains oriented with the crown pointing upward. English ornithologists refer to upside-down flight as "whiffling," a word surely derived from the whistling sound that accompanies it. The motor pattern is especially impressive when performed simultaneously by all the members of a large flock.

Upside-down flight does not exert a braking effect. Braking is apparently achieved by a particular wing posture that I have termed the bell posture (Figure 31). In this posture, the wing tips curve downward and in extreme cases can even point inward. The body axis is gradually raised and the feet stretched downward. Then, when the bird is close to the ground and its speed has slowed to the minimum for gliding, the feet are held against the sides of the tail fan and wing beating begins. The thrust generated by the wings can be oriented in any direction desired, including a slight backward impulse.

Figure 31: *The bell posture. In order to brake during fast flight, the goose twists its primaries and stretches its feet forward.*

Point Landing

Geese can easily reduce their speed to zero and land on a specific spot by flapping their wings (Figure 32). As we have already noted, wing flapping requires a great deal of muscular energy, and geese perform an exact-point landing only when it is really necessary. On a flat meadow, a goose will typically run forward a few steps after landing, and on open water it will brake even less. The kinetic energy is dissipated by water-skiing, which often covers a distance of several meters (Figure 33).

The most effective process of instrumental learning in geese is not the generation of a new motor pattern but the adjustment of an innate pattern. For example, we can observe exact-point landing in geese that chose to nest in trees. The restricted nest-building patterns of anatids do not permit them to construct a nest base in a tree, but, doubtless as a result of genetic programming, they show a strong inclination to take over large nests made by other birds. Their first attempts to land on a tree nest are often unsuccessful, but before long they hardly ever have to make a second attempt at landing, whatever the direction of the wind. Heinroth once observed some nest-seeking geese attempting to land in the

Figure 32: *In order to land on a specific spot, a goose must brake to a standstill in the air and alight vertically with its wings flapping.*

dense crowns of trees and, of course, tumbling through the branches when they failed to find a foothold. He placed nest baskets in the trees, and the geese, after numerous attempts, finally learned to land on them.

Escape Behavior

Escape behavior is a universal term referring to successful or unsuccessful attempts to get away from an apparently threatening object. The lowest intensity of escape behavior in geese is represented by vigilance. The goose stretches its neck vertically and holds its beak exactly horizontal (Figures 34 and 35). In this posture of pure vigilance, it is the eye that is held as high as possible, whereas in demonstrative display or threat it is the tip of the beak that is held the highest. A vigilant goose keeps a careful watch on the object that has

Figure 33: *When landing on water, a goose does not need to brake to a complete standstill by flapping its wings. The braking function is taken over by the feet meeting the water.*

Figure 34: *A gander showing vigilance.*

131

Figure 35: *A group of geese showing simultaneous vigilance.*

elicited this escape behavior and also may utter alarm calls (pages 171–176).

Male geese show vigilance more often than females do. In fact, certain high-ranking males serve as sentinel ganders, taking over a large part of the vigilance activity of the flock and grazing less frequently than the other geese. It is interesting to note that the same individuals that show the greatest courage when facing up to conspecifics also show the lowest escape threshold, that is, the greatest caution with respect to another species that might pose a threat. A vigilant goose has a greater tendency to flee, either by running away or by taking to the air. At all times after they are able to fly, geese respond to major alarm stimuli by flying off. I have observed escape on foot at maximum speed only when a particularly powerful escape stimulus, such as the eagle-warning call, has aroused the geese and there is a secure stretch of open water nearby.

Many mammals show similar high-intensity locomotor

patterns, particularly during vacuum activity or play. In other words, at their highest intensity the instinctive behavior patterns of locomotion show exactly the same form as during desperate attempts to escape. Any rider knows, for example, that damping a horse's locomotor drive may lead within a few days to an outbreak of pure locomotion in the form of bolting. At its highest intensity, the bolting is accompanied by such motor patterns as bucking and abrupt stalling with the neck held low, which had the original function of shaking off a large predator.

In bursts of high spirits, birds can fly upside down, perform loops, and so on. In our discussion of play diving a few pages back, we said that motor patterns observed in this context probably developed under the selection pressure of predator avoidance. That is almost certainly the case with the pattern of flying upside down.

When there is a high level of motivation to flee, the response of cringing may occur instead of active escape. In this, the bird dashes to the nearest cover or crouches down wherever it happens to be standing (Figure 36). In either case, the high degree of arousal of the sympathetic nervous system is apparent from the pronounced protrusion of the eyes. The tendency to crouch seemingly depends on the goose's hormonal condition, since even pinioned birds that will never be able to fly show a greater predisposition toward this behavior pattern during the molt.

Situations Evoking Escape Behavior

Throughout this section of our ethogram of the greylag goose, it is only natural that we have been saying more about innate motor patterns than about innate releasing mechanisms. Situations that evoke escape behavior, however, are an exception, because the releasing mechanisms are not hard to analyze through observation under natural conditions.

Like all higher vertebrates, greylag geese are afraid of anything that is unfamiliar to them. For instance, they must

Figure 36: *A gander crouching, or cringing, in an open area at the water's edge.*

overcome a strong inhibition whenever external circumstances, like an overpowering need to feed, oblige them to land
on unknown terrain. This requires great patience from a
human caretaker who is trying to induce them to land. It is
exasperating to have the geese, arriving with their feet outstretched and their wings set to brake, repeatedly break off
the landing attempt at the last moment and swoop upward
again. A large area of open water in unknown terrain has a
calming effect, as does the presence of a group of peacefully
feeding conspecifics. A familiar human is reassuring to hand-
reared geese. In this case, the human caretaker has the same
effect as an entire flock of geese.

Any object that appears suddenly, even on familiar terrain,
is a powerful stimulus evoking escape. It might be a car
emerging from a wood, a horseman, or even a conspecific
appearing unexpectedly at the edge of a wood. An object that
appears as a dark silhouette against the sky will evoke rest-

Figure 37: *One-eyed monitoring of an object high in the sky, probably.*

lessness and vigilance behavior (Figure 37). As Tinbergen and I showed in some experiments with dummies at Altenberg, a key feature of a flying predator in provoking escape behavior is slow movement (in terms of multiples of the length of the dummy).

As we later discovered, the shape of an object eliciting an escape response is less important that the frequency with which it appears. An object that appears regularly quickly loses its effect. We did not carry out any experiments with dummy predators on our geese at Grünau, but we know that herons, which fly over rarely, evoke a greater response than buzzards or ravens, which we see all the time. The extremely strong response of the geese to the rarely observed golden eagle, however, seems to call for a special explanation.

Intense alarm behavior, as a form of escape behavior

directed against nonflying predators, is evoked in greylag geese by any fur-covered object, such as a bundle of fox fur. The same response, of course, is shown to approaching dogs, especially if they have rufous fur. But the response rapidly wanes with habituation. In Seewiesen, we worried that the geese might be endangered by their habituation to my crossbred chow-chows, which were similar to foxes in appearance. But our fears were unfounded. The geese soon learned to accept my dogs as harmless, while continuing to show an undiminished fear response not only toward foxes but also toward unfamiliar chow-chows.

Greylag geese are particularly perturbed by anything moving underwater. When wading in front of a goose, one must never strike the goose's belly with an ankle, as this can elicit an intense escape response.

Preening and Stretching

Although it consists of dead material and not living cells, a bird's plumage is a marvelous and complex affair. It combines the functions of insulation and thermoregulation, the latter accomplished by raising the feathers to increase the layer of air surrounding the body without reducing the outer shield. Also, the plumage of many birds is almost completely impermeable to water, a somewhat puzzling phenomenon.

The special function of waterproofing has led to the evolution of a large number of special motor patterns that are as ancient as the plumage itself. The phylogenetic antiquity of the broad array of preening activities is obvious from the fact that they occur in many birds besides the greylag goose. Myra A. Mergler has studied these activities in our geese and written about them in a monograph. All birds with a breast keel (Carinatae) use essentially the same set of motor patterns in preening. These are innately coordinated movements that

Figure 38: *The quill of the feather is drawn through the beak along its undersurface.*

exist in pairs, for the left and right sides of the body. Some of them appear before the bird leaves the nest, and others emerge right afterward. It is a constant surprise to observe the self-evident skill with which a barely dried gosling per-forms its preening activities, as though it has already done so a thousand times.

Preening Activities

In most preening movements, the feather is held between the tips of the beak and is pulled through the beak from its base to its apex (Figure 38). Nibbling of the feather can also occur, especially if it is still growing and covered by its horny sheath, making the quill inaccessible. The quill of the feather is also carefully nibbled on the undersurface, again passing from base to apex. When a goose bends forward to preen its belly feathers, it is easy to draw the false conclusion that the preening movements are being performed against the grain. But here, the feathers are being grasped on the undersurface. As this is done, the tip of the beak is held against the bird's body, which of course directs the crown of the head toward the tail. This vital preening movement may look somewhat

Figure 39: *Oil is spread over the head by rhythmic scratching.*

undirected, but it is capable of straightening out feather vanes in which the hooks on neighboring barbs have become disconnected. This is done with a firm, secure movement that is somewhat like fastening a zipper. After preening the down feathers, the goose smooths them by stroking the lower side of its beak and the thoroughly oiled side of its head over the fluffed-up feathers on the chest and belly.

Two special motor patterns are used in spreading on the oil from the rump gland. The tail is moved into an extreme oblique position to the left or right of the body, and the upper tail coverts are spread as wide as possible. After this preparatory adjustment, the goose can remove oil from the rump gland with its beak, and this may be followed by a rhythmic scratching movement with the middle toe, which removes oil from the beak and spreads it over the head (Figure 39). Alternatively, the entire head may be rolled over the rump gland, in a peculiar rotation around the longitudinal axis that permits all parts of the head to come into contact with the secretion (Figure 40). The oil from the rump gland is then spread over the back and flanks by head rolling (Figure 41).

The static electricity generated by these and other preening movements appears to be important for the waterproofing

138

Figure 40: *The head is anointed with*   **Figure 41:** *Head rolling.*
oil by rubbing it on the rump gland.

of the plumage. Indeed, according to experiments conducted by Ernst M. Lang in Basel, it is more important than the oiling of the feathers. Lang removed the rump gland from adult mallards and released the birds on a large pond in the zoo. He observed no difference in them in comparison with a control group.

Several things support the assumption that the electric charge is important. In young goslings, waterproofing of their plumage is clearly linked to the precise spacing between the ends of their down feathers. Underwater photographs of diving ducks often show that any individual feathers that are out of place in the orderly arrangement are wet. We can see this because in a photograph such feathers are in sharp focus, while the other feathers are covered with a silvery layer of air.

Another indicator of the importance of the electric charge is that those waterfowl that remain almost constantly on the water, such as grebes (Podicipedidae), preen themselves almost without interruption. In fact, some observers have suspected that these birds must be riddled with parasites because they are always scratching themselves. Anyone who has watched a great crested grebe on our lakes is familiar with the sudden flash of the broad white belly as the swimming bird rolls far over to one side in order to tackle the belly feathers with its beak. This type of care of the plumage seems to become more necessary as contact with the water increases. Paradoxically, waterfowl that swim almost constantly are the quickest to lose the waterproofing of their plumage. When small grebes are being transported, a reduction in waterproofing is apparent after only a few hours. Diving ducks also have trouble coping with a lengthy dry transportation, but it has no adverse effect on mallards or geese, so long as the surface of the plumage remains free of dirt.

Figures 42, 43, and 44 will serve as further description of the wide range of instinctive motor patterns involved in preening.

Bathing Activities

The instinctive motor patterns that serve to wet the plumage are apparently just as ancient as the preening activities. Such distantly related groups of birds as anatids and songbirds have clearly homologous bathing patterns. Wallace Craig, in his studies of the ring dove, found that drinking is related to bathing activity. When an inexperienced dove encounters a large water surface, its first response is to shake its beak, which splashes water around. In an older bird, such beak shaking is the first motor pattern to occur prior to bathing. In a young dove, however, contact between the beak and the water elicits the first drinking response. This transition from an innate bathing pattern to learned recognition of a water surface seems to take place in most, if not all, doves.

Inexperienced anatids respond to their first contacts with water by drinking. If bathing follows, as it usually does, the bird starts by shaking its head so that the beak and the front part of the head touch the water. Then the head is quickly dipped deep in the water and turned sideways in such a way that it will throw a jet of water onto the back when it is flipped upward. At the same time, the bird begins to move its wings up and down from the shoulder joint, but without opening the wings at the other joints. Supplementing the scooping movements of the head, this movement serves to throw more water onto the back. As the intensity of the wing movements increases, however, the head dipping ceases. Now the bird rears up somewhat and holds its shoulders tilted to one side, causing the wing on the opposite side to pass over the back and dip into the water up to the elbow (Figures 45 and 46). The movement becomes faster and water splashes high in the air. The distinctive noise of this splashing can be heard from a long distance away.

As the bathing reaches an even higher intensity, the bird suddenly tucks its neck into its shoulders and stretches its head far forward. The tail is turned almost as far forward toward the belly, and the wings—still being moved only from the shoulder joints—are plunged over the head into the water to produce a veritable shower. Ducks and most of the other birds I am familiar with keep their balance during this maneuver and at once resume their previous position. Geese, however, tip forward (Figure 47) and then, surprisingly, swim a little way on their backs, with the belly uppermost and the legs treading the air (Figure 48). Heinroth has reported seeing one of two young greylag geese react with great alarm the first time its sibling adopted this position.

As their bathing arousal declines, both ducks and geese often float on the water for several seconds with their head feathers fully ruffled, the neck tucked in, and looking as if they are yawning. In fact, they are stretching the jaw musculature, but without the respiratory and other stretching movements that accompany yawning in mammals (including humans) and in many reptiles. To avoid confusion with genuine yawning, we use the term hyoid adjustment for this motor pattern (Figure 49).

Figures 42, 43, and 44: *A selection of the wide variety of preening patterns.*

Figures 45 and 46: *In bathing, both wings are dipped into the water to splash it upward.*

Figures 47 and 48: *The goose rolls over and swims on its back for a moment.*

Figure 49: *This is not real yawning, but simply a stretching of the hyoid musculature.*

Drying and Adjusting the Wings and Wing Shaking

I include among bathing activities those motor patterns that permit rapid drying of the wing feathers, especially the primaries, after the bird has left the water. Anatids spread their wings in alternation, stretching them so that the primaries fan out, and then fold them back into the resting

146

Figure 50: *Wing shaking. The ripples on the water show that the thrust from the wing movements is directed straight upward.*

position. Passerines, even small species, beat the air with their primaries while drying themselves; the speed of the movement produces a buzzing sound.

Adjusting the wings is the motor pattern by which the goose alternately raises its dry wings from the shoulder joint and pushes them back into the wing pockets with a rubbing movement. I have seen the same motor pattern in storks that I had temporarily prevented, using feather clamps, from flying.

In the motor pattern Heinroth called wing shaking, the goose rears up as high as it can and performs noisy wing movements that have exactly the same coordination pattern as wing flapping (Figure 50). The goose will often stretch up onto the tips of its toes to perform this pattern. Indeed, a young goose just before fledging may lose contact with the ground completely.

To conclude this section on care of the plumage, we will

discuss a number of motor patterns that fall under the label of comfort activities.

Body Shaking

I have observed a brief head-shaking movement in reptiles, mainly in lizards, but it is a single action and is not rhythmically repeated. Its homology with the body-shaking activities of birds is therefore dubious.

When a bird shakes itself, it begins by fluffing its feathers from back to front. That is, the fluffing begins with the down feathers of the lower back and moves forward, accompanied by rhythmic to-and-fro movements of the body along its longitudinal axis that start with the lateral wiggling of the tail and end with the shaking of the outstretched head (shake-stretching). The shaking movement is performed without any movement of the wings (Figure 51). In geese, the coordination of the movement seems to be confined to the spinal cord, for Heinroth has reported that the sequence can take place even in a goose whose head has been severed. With a passerine bird, by contrast, the entire body goes limp after such an operation.

Certain components of this shaking pattern, which involves the entire body, have taken on a life of their own. While maintaining a normal body posture, both geese and ducks can shake just the tail from side to side. This is the commonest displacement activity for both groups of birds and is performed when, for example, a minor problem is encountered in climbing out of the water. When this happens, a duck will merely waggle its tail. Tail shaking is particularly conspicuous when a bird defecates while lying down, because then the tail is slightly raised.

A goose can also shake its head from side to side, and this movement can grade into intermediate forms of the shake-stretching movement. Finally, lateral shaking of the wings has also developed into an independent pattern in geese. It is

148

Figure 51: *In shake-stretching, the twisting movement of the head and neck about the longitudinal axis is so fast that the bird's eye cannot be seen. (This bird is a hybrid between a swan goose and a white-fronted goose.)*

done not only when there is something on the wing, like water, but also in exaggerated form as a displacement activity. We call this latter pattern of noisy wing shaking, which occurs in a threat context, wing rattling (Figure 52).

Stretching

Anyone who has owned a dog is familiar with the two stretching movements of mammals. In one movement, the forelimbs are stretched far in front of the body with the chest close to the ground and the hindlimbs upright. This is commonly followed by the second movement, the backward stretching of the hindlimbs with the hindquarters lowered. There is one ancient Oriental illustration of a lion that has

Figure 52: *Wing rattling, a shaking of the wings that has developed into an exaggerated, noisy display. It is a common form of conflict behavior in threat contexts.*

been generally interpreted as showing an injury of the spinal cord; in my opinion it only shows the second form of stretching behavior.

The stretching movements of birds are remarkably similar to the mammalian patterns, except that it is obviously impossible for both legs to be off the ground at the same time. The backward stretching of one leg is typically accompanied by the stretching of the wing on the same side of the body, with the foot brought close to the fanned-out primaries (Figure 53). (There is a misleading statement in the literature that a swan can straighten out its primary feathers with this movement.) Greylag goslings lose their balance when they try to perform such a stretch.

Another stretching movement of adult geese is performed with both wings or both legs simultaneously. The legs are stretched as far downward as possible from the knees and ankles, which conspicuously raises the bird's back. Frequently, the wings are raised from the shoulder joints at the same

Figure 53: *Wing stretching.*

Figure 54: *Simultaneous upward stretching of the wings.*

151

Figure 55: *The typical body posture during heavy rain or hail.*

time, with the elbow and wrist joints sharply bent (Figure 54). The two movements occasionally occur in isolation. Young geese stretch both legs backward while lying on their bellies.

As far as I am aware, these stretching movements are found in all birds, or at least in all birds with a breast keel (Carinatae). Stretching from the ankle joints is particularly conspicuous in passerines, in which the tarsus is almost horizontal when the bird is in the normal sitting posture. The simultaneous extension of the ankles and the wings bears a remarkable similarity to the corresponding movement in mammals, especially since both birds and mammals are able to extend the neck and lower the shoulders at the same time. But such a full coordination of stretching movements is not obligatory.

Scratching

All four-limbed vertebrates exhibit motor patterns that let them remove particles adhering to the head. Rhythmically repeated scratching is presumably a convergent adaptation for dealing with long appendages on the skin (hair or feathers), which are combed in a similar fashion and apparently receive a static electric charge. Of course, when an adult bird is scratching its head with one foot it must stand on the other. Yet the pattern emerges in many young birds well before they are able to stand, in both altricial and precocial species.

A particular form of scratching, believed to be common to all ducks and geese (Anatidae) and therefore found in greylag geese, has been described in the section on preening activities. As we noted there, in this special form of scratching the first movement is directed at the tip of the beak, after the beak has been used to remove oil from the rump gland, and the oil is spread over the plumage.

The physiological nature of all comfort activities means that of necessity they frequently appear as displacement patterns. For this reason it is worth noting that scratching, surely the commonest displacement activity in humans, never occurs as such in birds.

Feeding Activities

Every bird has a beak—skeletal components of the upper and lower jaw joined to form an acute angle. In all species with which I am familiar, the beak's sharp tip is protected by a horny sheath. At an embryonic stage, the beak also bears the egg tooth, a structure composed of histological elements of a true tooth. The egg tooth drops off shortly after hatching.

Pecking

All the birds I have ever observed use the same motor pattern in feeding. The beak is aimed at an object on the ground in sight of both eyes, and usually an audible sound is produced (Figure 56). The food item is grasped between the upper tip of the beak and either the tip of the tongue or the lower beak

Figure 56: *A gosling pecking at an object fixated with both eyes.*

tip, and it is then transferred to the mouth cavity with a jerk of the head. Chicks use the upper beak tip and the tongue, while goslings use the upper and lower beak tips as pincers. The movement by which the food is tossed into the mouth cavity is what is described later as shoveling. It is particularly well developed in birds with a very short tongue, like some starlings and the toucans. A toucan's beak slit has exactly the same parabola curve that the food item follows on its way from the tip of the beak to the mouth cavity. With this type of feeding behavior, the bird does not need to open its beak very wide.

Inexperienced greylag goslings peck at all kinds of objects, without showing any obvious color preference. It is only necessary for there to be a definite contrast between the object and its background. Although a learning process rapidly leads the goslings to prefer green, they continue to be attracted to sparse patches of grass at the edge of a path because, unlike the grass in a lush meadow, the blades contrast starkly with the dirt background.

Plucking

Soon after goslings have learned to peck at green objects, a motor pattern emerges in which they grasp a blade of grass between the upper and lower jaws while pulling it up with a powerful backward movement of the head and neck. At lower levels of intensity, the neck musculature supplies the power for the movement. At maximum intensity, as when a goose is pulling out deeply rooted plants, the neck is stretched out in front, with the jaws clamped together, and the goose throws its body backward with all its might. Reeds may have tough roots, but the force of the movement is especially apparent to a human observer when a goose uses this motor pattern to pull off a tablecloth and all the dishes on it.

Drawing Through and Slicing

A motor pattern mainly used during grazing on broad-leaved plants and reeds enables a goose to slice a sharp-edged leaf by drawing it through the beak from the base to the tip (Figure 57). Seeds are gleaned from the heads of grass with what I regard as essentially the same movement. This seems to be a very profitable pattern, for one often observes a goose rearing up beside a grass stem in order to get a good hold on it. If the goose meets any kind of projecting obstacle while drawing a pliable but resilient object through its beak (like the snap fasteners on a raincoat), the movement will give way to plucking.

Shoveling

Another feeding pattern, first observed in the greylag goose by Christa Walter, is what we refer to as shoveling. This motor pattern serves to transfer food from the front part of the beak to the throat. As already noted, it is present in birds

Figure 57: *Oblique slicing.*

with a short tongue, but it is also shown by anatids, in which specialization of the tongue for sieving blocks other functions. This motor pattern's most conspicuous performance is when a goose takes a beakful of seeds from a pile and swallows them.

The same pattern is used when an object is likely to slip out of the beak. For example, if a goose has bitten off a large chunk of turnip, it will use the shoveling pattern in order not to lose a morsel—particularly if other geese are trying to steal it. Since, under natural conditions, a greylag goose will hardly ever have a chance to push its beak into a pile of seeds or something similar, shoveling is presumably used mostly to deal with large morsels like the turnip.

Nibbling

As soon as young goslings begin to peck, they also nibble at various objects. Nibbling is a motor pattern that appears to have an explorative character, to the extent that it is applied to unfamiliar objects and can at any time give way to a different pattern. Inquisitive nibbling is the only motor pattern of the greylag goose that I regard as having an explorative character and even, to some degree, a playful quality. In some goslings reared by a human caretaker who withheld greeting, nibbling was extinguished, in the same way as exploratory behavior was extinguished in the rhesus monkeys that Harry Harlow reared under conditions of social deprivation.

Nibbling at familiar humans is of special significance with hand-reared greylag geese. If the person has some feature that elicits this motor pattern, such as shoelaces or long hair, the young geese can become extremely vexatious. It seems impossible to train goslings not to nibble, and it certainly cannot be done without permanently reducing their attachment (Figures 58 and 59).

Upending

Upending is one more motor pattern the greylag geese use in feeding. In upending, the tip of the beak, in marked contrast to the sieving movement, is kept in the same place as the base of the beak moves from side to side and the head presses forward in a boring motion. The neck is usually directed downward at a right angle to the water surface, so that the tip of the beak bores into the bottom (Figures 60 and 61).

The primary function of the motor pattern is doubtless the collection of starch-containing roots and other nutritious food. Occasionally, adult geese have been observed gathering *Elodea* from the bottom of a pond and letting their goslings eat it. As a rule, however, a goose has nothing visible in its beak when it resurfaces after an extended bout of upending. Sometimes the movements of the legs and the rear of the body indicate that the bird is using tugging movements in trying to overcome a strong resistance.

Figures 58 and 59: *Nibbling as exploratory behavior.*

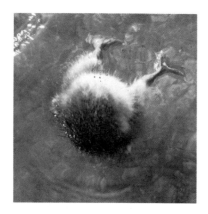

Figures 60 and 61: *Even when only a few days old, goslings show upending just like that of adults.*

Upending is completely instinctive, as we know from simple observation and from the results of an equally simple experiment. My goose pond in Altenberg contained no edible plants and had a clay bottom covered with pebbles. The geese nevertheless spent a great deal of time upending, showing a preference for water just deep enough for them to reach the bottom. Whenever they stopped upending, I could induce them to start again by throwing a few handfuls of corn into the water. Obviously, they were upending in order to eat. But if I kept throwing corn into the pond until they could no longer be induced to upend, they would still eat a handful of corn placed on the bank of the pond.

This showed that two independent motivations could elicit upending. According to their relative strength, sometimes one and sometimes the other will determine the outcome, in line with the principles set out by Paul Leyhausen in his discussion of the relative hierarchy of motivation.

Sieving

Sieving is the motor pattern through which anatids obtain a large part of their food. The beak is slightly opened, and the

Figure 62: *Adult geese mainly sieve their food (water lentils, mosquito larvae) from the surface of the water.*

tongue moves rapidly to and fro in the mouth cavity like a piston. The movements suck in water near the tip of the beak, along with any particles suspended in the water, and expels the water to the side. The water passes through a filter formed by the denticles on the tongue and beak, which retains any small bits of food (Figure 62).

Modifications in the form of the beak occurred repeatedly during the evolution of the anatids, as the goose group of the Lamellirostres converted the lamellae of the original sieving beak until they could bite off pieces of plants. In geese, the only vestiges of the ancestral sieving apparatus are the blunt denticles on the upper beak and tongue and the corresponding tubercles on the lower beak. The sieving movement of the tongue was likewise reduced. A clearly developed form of the sieving pattern is seen only in very young goslings, which use it to gather small stones for the gizzard. Young goslings show an obvious preference for small puddles standing on paths of sand or pebbles (Figure 63). In fact, I cannot imagine

Figure 63: *A gosling collects the small stones it needs for its gizzard by sieving.*

where else along the fast-flowing River Alm, with its bed of large pebbles, our goslings would have found stones small enough for their gizzards.

Motor Displays and Vocalizations

Most displays of the greylag goose have a simple type of central nervous system coordination, and conspecifics under-stand them as signals. All are important, to varying degrees, in the social system for which they provide the basis, and the reader needs to know this "vocabulary."

The Distress Call

The call that we have labeled lost piping, and have briefly referred to as crying, is not only the first vocalization to be heard from a gosling while it is still in the unhatched egg; it

161

is also the source from which several other vocalizations derive, in a largely epigenetic development. Helga Mamblona-Fischer describes the first distress call this way: "It is a high-pitched, monosyllabic call. It is produced [before hatching] when the egg is markedly cooled or when the gosling encounters difficulty in hatching, for example if the egg membrane dries out. Goslings that cry frequently before hatching usually die." Obviously, then, the call is not restricted to occasions when the gosling is lost. "When a gosling utters a distress call from the egg, the mother peers beneath her belly and moves to and fro, just as she does later on when she makes a recently hatched gosling cry by stepping on it. Subsequently, the mother sometimes performs egg-rolling movements. Turkey hens as foster mothers sometimes squash goslings to death because they do not respond to their distress calls. This never happens with the much more vulnerable pheasant chicks, because their distress call is similar to that of the turkey."

The distress call obviously serves the function of attracting the attention of the parent in situations where the gosling is threatened (see Plate I). Goslings begin to cry when a problem arises during hatching or if they fall too far behind the leading parents and feel they have been abandoned. They also cry whenever they are hungry or thirsty, in which case they may even take over the leading role. If a flock of geese is resting near the bank and the goslings feel the need to drink, they may run energetically away from the parents and toward the water, crying loudly.

It is surely reasonable to regard the distress call as an expression of acute discomfort. I am inclined to agree with Hans Volkert's conjecture that the experiences of pleasure and displeasure arose at a point in evolution when it became desirable to develop a common signal for internal and external stimulus situations having positive and negative conditioning effects. The best phrase I can think of to cover all the situations in which the distress call of goslings is evoked is this: anything that is a threat to the survival of the individual.

Figure 64: *The voice of the young goose breaks at the time of fledging. Nevertheless, like older geese, they can revert to infantile crying.*

We can probably learn a great deal from the situations that elicit distress calling in older geese (Figure 64). For example, one November two ganders that had hatched earlier in the year landed on thin ice and broke through. They first attempted to climb through the hole and onto the edge of the ice. After failing several times, and without even trying to fly away, they started to cry loudly. The calls had a strange sound, since their voices had already broken. In between the distress calls, they also uttered lamentation calls.

A second incident was more comical because of its close analogy to human behavior. A tame greater snow goose (*Anser caerulescens atlanticus*), appropriately called Little Princess, was trained to come to me because I always carried a handful of wheat, which I offered her every day the first time we met. One day the box of wheat was empty, and so I took a handful of oats. As usual, Little Princess approached me joyfully. As soon as she spotted the oats in my hand, however, she reared her head and began to cry loudly in the deep voice typical of adults of her species. Anyone who has had a similar experience surely shares my deep conviction that higher animals do have a capacity for subjective feelings.

The remarkable feature of the distress call is its versatility. The same simple motor pattern can express a great variety of motivational states, ranging from disappointment to a reaction to bad food to a response to an attack or a defeat. This signal's uniformity in many different social situations is in stark contrast to the great variety of displays represented by the contact call. Interestingly, the contact call and the distress call stand at the opposite poles of attraction and aversion, alliance and antagonism.

The Trilling Call

After an egg has cooled enough that the unhatched gosling has uttered a distress call and the mother has warmed it again, a trilling call will be heard. The same sequence of vocalizations is produced after hatching when a gosling has become cold and has been warmed again, by creeping under the mother or by an artificial source of heat. The trilling is also heard when goslings scuffle with each other in their sleep. This call persists at least until fledging, and longer in some individuals.

The Good-Taste Call

The good-taste call, a monosyllabic vocalization, is so seldom heard that I would be inclined to omit it except for the fact that it occurs in various related precocial birds, indicating that it has some importance. When domestic chicks begin to feed shortly after hatching, the short *pip* they utter when pecking is slightly modified by a deeper overtone that transforms it into *pweet-pweet*. This call is a sign that the chick is really feeding. Mallard ducklings have a similar call, closely resembling that of domestic chicks in both the releasing situation and the vocal pattern. The ducklings utter it when ingesting their favorite foods. In all probability, it is homologous with the food-summoning call of adult chickens.

I have heard geese utter this call on only a few occasions, and every time, it came from young goslings, a few days or weeks old, as they fed on hatching gnat pupae on the water. I was familiar with the sound of the call and the accompanying behavior from ducks, but I had not known that it also occurs in geese. In fact, I have never observed geese eating any other animal food so greedily. The call is apparently linked to psychological and physiological conditions that occur rarely.

The Lamentation Call

The lamentation call is also monosyllabic. It is a remarkable but unexplained fact that expressions of displeasure or lamentation have a similar quality not only in humans and many other mammals but also in the subject of our study, the greylag goose. A dog whining at the door, a lost gosling wandering in search of contact, a human baby or a young beaver whose feeding is stopped, and an adult human whose work has been interrupted all utter a vocalization that begins on a high frequency, rises briefly, and then sinks.

Even very small goslings utter this monosyllabic, drawn-out, pure call. They produce it in the same stimulus situations as those that elicit lost piping, if the situations exceed a certain intensity but last only a short while. The head is raised high, the feathers on the head and neck are flattened, the eyes bulge outward, and the beak is held horizontally. Mixtures of the distress call and the lamentation call have been noted, but I have never heard a long sequence of repeated lamentation calls such as occurs with distress calls and distance calls.

The lamentation call is most frequently uttered when a goose that has been beaten in a fight is attempting to flee but has been seized by its opponent and is being held firmly by the feathers on its back. Heinroth has said he never heard the lamentation call except from geese that were still associating with their parents. But we believe this reflects a confusion with the distance call, which is initially monosyllabic and at that stage difficult to distinguish from the lamentation call.

In Königsberg, I once heard a young greylag goose that was flying in circles over the town in thick fog continuously utter a drawn-out, pitiful call that I interpreted as a lamentation call. A detailed study of the similarity between the lamentation call and the juvenile distance call, however, has led me to believe that it was more likely the latter I had heard. Since both the lamentation call and the distance call are definitely derived from the distress call in their ontogeny, it is an easy mistake to make.

Pestering and the Departure Call

There is understandably a close connection between discomfort caused by an external situation and the intention to move elsewhere. It is therefore not surprising to find that in many animal species the call expressing discomfort also signals the intention to depart. Pestering, a call or sequence of calls of falling frequency, has a complaining quality, whether uttered by a bird or by a mammal.

The so-called singing of the domestic hen prior to egg laying, which expresses the need for a nest, and the discomfort-quacking of the mallard—"I want to go somewhere else"— have a similar ring. And the egg-laying vocalization of the hen pheasant sounds just like the distress call of a chick.

The nonmodulated departure call of the adult goose and the modulated pestering call of the immature bird share a common eliciting context. When hungry adult geese crowd around a human caretaker begging for food during the winter, say, they rely on the departure call. The call has a general air of discomfort and is given intensively, sometimes accompanied by beak shaking.

Because of the connection between discomfort and the intention to depart, I at first looked for a discomfort call of the greylag goose that could be linked to the contact call (page 193). The basic question is whether the melody of the lamentation call can be interpreted as a series of contact calls that have taken on a complaining character, appropriately referred to as griping. Young greylag geese produce certain sound sequences that not only are readily understandable in this sense but also, significantly, can lead directly to distress calls. Pestering is closely related to distress-calling, at least in its motivation. Young goslings will sometimes coax their mother to come with them by uttering the distress call and the pestering departure call, one after the other or mixed. The possibility of expressing discomfort with sequences of the contact call apparently disappears gradually as the voice breaks. The pestering call, a plaintive melody composed of contact calls, is then replaced by the departure call proper.

The departure call is also derived from the contact call, with the difference that, unlike the call just described, every element has exactly the same frequency. An adult goose expresses dissatisfaction and the intention to walk or fly away by interposing monotonous departure calls between contact calls, with the sequences of departure calls becoming longer until eventually they predominate.

Besides its monotony, this specialized departure call of the greylag goose is characterized by its staccato structure and by the disappearance of the overtones that are clearly heard in ordinary cackling. But the monotony of the call appears to be its primary feature, as we realized from listening to our bean geese preparing to leave the Auingerhof to fly to their sleeping site on Lake Alm. I was asked to interpret the calls of these geese and without hesitation identified them as departure calls, even though the features of change in pitch and staccato performance are lacking in these geese.

At a high intensity, the monotonous departure call is accompanied by sideways shaking of the head, a movement that doubtless arose through the ritualization of a displacement activity. It is easy to induce geese, particularly those imprinted to follow humans, to take to the air by simulating the departure call and imitating with a hand the shaking of the head. Interestingly, young geese interpret the hand movements of the human foster parent the same as they do the head movements of their natural parents. It is not even necessary for the arm to be stretched out and the hand held at right angles to the forearm. In fact, the hand movements are more effective when they are performed at the level of the goose's head. Young shelducks (*Tadorna tadorna*) also can correctly interpret such hand imitations of head movements.

The geese's preparations for takeoff actually permit one to predict whether they are really going to take off or not. I have been able to do so for some time, but it took a great deal of observation before I could identify the right criterion: it is the speed with which the different phases of increasing excitation follow each other. One can extrapolate the curve of the growing excitation, and if the individual phases occur rapidly in the middle part of the trajectory, it is certain that the excitation will increase still further. This increase in the action-specific excitation for takeoff clearly tells the observer that the bird totally lacks any capacity for voluntary decision making. If the goose could speak, it would not say "I want to fly" but "I am about to be flown." It is comparable to the state of a person afflicted by a growing impulse to sneeze: the individual seeks through all the available voluntary means to reach the threshold that will elicit the explosion.

The departure displays and vocalizations I have been describing not only are highly infectious but also have a strong feedback effect on the transmitting animal. If the individuals receiving the signals do not respond and fail to show any signs of motivation for takeoff, there is a demonstrably strong inhibitory effect on the transmitting individual. Reciprocal influences of this kind are more pronounced among family members than among strangers, and naturally are the strongest between the partners of a pair.

Figure 65: *The distance call. In contrast to the rolling call, the hyoid bone is not lowered, and therefore the neck does not bulge.*

The Distance Call

An adult goose that feels it has been abandoned will stretch its neck high, with the beak horizontal, and utter a loud trumpeting call of one to five syllables (Figure 65). Most typically, this distance call has three syllables, the first one accentuated and in a somewhat higher frequency than the others. The call sounds something like *gig-gag-gag*. It is heard, for example, when a goose becomes separated from its group or when offspring are present and the partner is not close by. It is always elicited when a goose's partner appears high in the sky after an absence. Then, there is never any doubt that it is the mate uttering the call.

In its juvenile form, the distance call closely resembles the lamentation call and is similarly derived from the distress call.

Habituated geese respond very well to a human imitation of the distance call and will reply to it from some distance away. In all probability, however, this highly specific response is a learned one. Small goslings become frightened when a parent abruptly utters a distance call nearby, and they respond in the same way as to an alarm call.

Distance calls show clear individual variations, and one can recognize particular geese by them from some distance away. And geese also are perfectly capable of distinguishing individual human voices. Whenever we call our geese from far away, we have to ask any guests who may be present not to join in the calls, as the geese are greatly frightened by unfamiliar human voices.

If we call a resting flock of geese over a long distance, they sometimes respond with a low-pitched call that we refer to as the somber distance call. After a few seconds, the normal polysyllabic distance calls are mixed in, and rolling vocalizations in the form of the puzzle call may also be included. As far as we can tell, this call is usually directed toward other greylag geese, and we are pleased when we are able to elicit it. On the other hand, we are not at all pleased to have our call answered with individual short alarm calls from the distant geese. This means that the geese have recognized the observer not as a conspecific but as a possible source of danger.

Because there are always several geese calling together, we cannot say whether there are any transitional forms linking the monosyllabic somber distance call with the ordinary distance call and the rolling call. Ordinary distance calls are repeated indefinitely, but it seems that the bird responds to a given stimulus only once with the monosyllabic distance call. The somber distance call passes once like a wave over a flock of resting geese and is then succeeded by other vocalizations.

The Hoarse Call

We know little about the meaning or cause of the hoarse call. This vocalization is an unvoiced, monosyllabic breathy sound that is uttered with the beak half open. Possibly it is an ordinary contact call in which the vocal organ is barely inflated, so that no proper sound is produced.

It is most often given by the female goose immediately after the hatching of her offspring, but it also occurs earlier, when infantile vocalizations are heard from the egg. A female goose will also give the hoarse call when, for example, she suddenly sees an egg lying in the grass. Hand-reared geese of both sexes utter the hoarse call in front of the food bucket before they start to feed or when a human hand is held right in front of their eyes.

A goose can occasionally be observed uttering the hoarse call while gazing at the ground, even though no object is visible. The goose stares at the spot with its eyes bulging and approaches as if spellbound while at the same time shrinking back in fear. The conflict between attraction for and fear of a certain spot can be clearly seen, but we do not know the reason for the response.

The Alarm Calls

Gog and Gig-gog

The greylag goose possesses three distinct alarm calls, each of which evokes a specific response from conspecifics. The commonest of these, the short alarm call, is best written as *gog*. It usually indicates mild disquiet and arouses the attention of conspecifics, carrying the message "Watch out, everybody!" The *gog* call is often uttered by a goose standing in an extreme vigilance posture with the head held high (Plate IV-1), but it can also be given from a lying position when the disturbing stimulus is not strong enough to make the goose get up. The characteristic feature of this vocalization is the orientation toward the releasing object. In disyllabic form, the call is given at two different pitches and sounds like *gig-gog*. The *gig-gog* call is commonly given by two geese (usually ganders) when they are alarm calling in alternation. Ganders produce the call more often than female geese do, and it is most likely to be uttered by the ganders serving as sentinels.

Gog calls usually precede the evening takeoff of the flock. I used to believe that their frequency was related to the general anxiety occasioned by the approach of nightfall, as is the case with the evening alarm calling of blackbirds. It seemed a reasonable assumption, because I first noticed the call as the flock was preparing one evening to fly about eight kilometers from the institute to the sleeping site on Lake Alm. Only later did I realize that there is no correlation whatsoever between the frequency of the *gog* call and the time of day.

The *gog* call also accompanies a behavior pattern that is known as mobbing.

Mobbing

Use of the word mobbing should not make anyone think that the process it describes has anything to do with the emotion of hatred. Curiously, mobbing animals appear to have no fear of the predator toward which the behavior is directed. Small birds mob owls, and ducks mob foxes or fox dummies. Swallows, white wagtails, and other agile birds seemingly have no need to fear raptors flying in the open. The term mobbing neatly expresses the way a group of weaker organisms can join forces to drive away a stronger adversary, which is indeed what the mobbing behavior of animals accomplishes. The activity is well known in birds, but it also occurs in similar fashion in bony fish. In nearly all cases, mobbing involves a sham attack on the adversary. It is highly exceptional for the potential prey to make an actual attack on the predator. The survival value of mobbing apparently comes about when a group of organisms thoroughly disrupt a predator's hunting by running or swimming behind it and loudly advertising its position. Long-lived animals with a well-developed learning capacity may even come to recognize that certain places tend to be frequented by predators. With their repeated *gog* or *gig-gog* calls, geese can signal that "there are often foxes here."

When a predator such as a fox is close to the water's edge,

so that geese or ducks can follow it as they swim, they do so eagerly and often approach alarmingly close. This well-known following response has been exploited in the trapping of waterfowl with Dutch duck decoys. (The English word *decoy* is derived from the Dutch "eendekoj.") A well-trained small dog, often wearing a fox fur, plays the role of the predator. Various duck decoys are used by ornithologists today to catch waterfowl for ringing, especially in England.

The *gog* call is the least selective of the alarm calls in its elicitation. The focus of the *gog* call and the associated mobbing response is hardly ever a rival conspecific. More typically, it is elicited by an external danger of some kind or by a conspecific that does not fit the expected pattern (because of illness, say) and therefore does not have the character of a rival. Even a familiar caretaker can elicit mobbing (Plate V-1) by behaving in an offensive manner, as when catching and ringing a goose. The *gog* call can be elicited by a small predator swimming in the water (Plate V-2) or by another greylag goose that has been sedated. We observed the latter response when we attempted to catch unringed geese by feeding them doctored pieces of bread. In such cases, the *gog* call can attract conspecifics and lead to a major escalation of the mobbing response. It can even induce the flock to take off, with the undesirable result that the sedated bird is also carried along. On one occasion, we saw an escalation of this kind lead to the production of individual eagle-warning calls. I have never heard a bird with this kind of behavioral disruption give the soft, or serious, alarm call, although that would probably arrest the mobbing response.

In the interests of completeness, it should be said that some ethologists refer to any high-intensity physical attack by a greylag goose as mobbing.

The Soft, or Serious, Alarm Call

The alarm call proper, more aptly termed the frightening call, is a short nasal *gang*. Heinroth recounts that in his youth, especially before his voice changed, he was able to imitate this call so well that he could drive flocks of geese to distraction, since they usually responded by charging off in fright toward the nearest body of water. This alarm call is often uttered very softly, especially when a pair is leading offspring and one parent discovers something suspicious.

At a low intensity, the call does not appear to be very frightening to the birds, and they do not always flee. They are in a state of full alert, however, and ready to flee at any moment. The soft alarm call is taken more seriously than the *gog* call, to the extent that silence falls immediately and the geese begin to look around intensively. This often lasts several minutes, until older geese abruptly give the all-clear by cackling.

When a flock of geese is surprised on open terrain, far from cover or from a body of water, often the entire flock will take off and fly rapidly skyward. Apparently the geese are safest from their natural aerial predators when they have plenty of airspace below them in which to perform their marvelous aerobatics, such as flying upside down and in zigzags. Molting geese that are unable to fly and geese leading small goslings respond to the first alarm call by heading for cover or for the nearest water.

In goslings whose following response is imprinted on a human or on a dummy, a wide variety of sounds can elicit the escape response. If the soft warning call is given through a loudspeaker built into a dummy, the goslings will dart beneath the dummy and press themselves flat on the ground. The response is not very selective, however. On one occasion, my coworkers connected a radio playing loud dance music to a dummy that I had attached to a moving bamboo pole. All the goslings responded to the loud music exactly the same as they did to alarm calls, and it was strange to see them fleeing toward the frightening stimulus rather than away from it.

When, in my first experiment, I uttered an alarm call standing up, the goslings did not gather close to me, but formed a dense clump some centimeters away. Apparently, the same mechanism was at work as that by which they maintain their distance from a leading parent. They judge the distance according to the height of the parent's outline.

Responses to imitations of the alarm call wane in a remarkably short time. Robert Hinde has made this observation in eliciting alarm calls and mobbing behavior from chaffinches with a dummy small owl. We do not know why the response, which is so important for survival, fades, but perhaps the monotony of the experimental conditions is in some way responsible.

The Eagle-Alarm Call

We have seldom heard the eagle-alarm call, and then usually on one of the rare appearances of a young eagle in the Alm Valley. The call is quite similar to the soft, or serious, alarm call, but is extremely loud and can be heard from a long distance away. All geese take this alarm call seriously and respond by fleeing to a large body of water or into thick cover. We also heard the eagle-alarm call once when a coworker was banding a greylag goose; a large flock of geese gathered around and mobbed her.

The response of greylag geese to a golden eagle is remarkable because of its selectivity. The geese give no response at all to the many buzzards that circle over the Alm Valley, which means they can distinguish the complexity of the eagle silhouette from the outline of the buzzard, which is not so very different. However, the geese uttered the specific eagle-alarm call the first few times they spotted the herons from the wildlife park up in the sky, although the response waned rapidly. Real eagles, which occasionally visit our valley from the Dachstein range, presumably appear too infrequently for any such habituation to occur.

As has already been mentioned, an extreme summation of the arousal elicitation of the common *gog* call can lead to the production of the eagle-alarm call.

Hissing

We will round off our discussion of alarm calls with a brief note about hissing, a nonsonantic sound produced by the pronounced protrusion of the hyoid apparatus accompanied by a powerful exhalation of air (Plate IV-2). Hissing is probably the oldest display sound to have developed in air-breathing vertebrates. Many species that puff themselves up during defensive behavior hiss when the air is released. Hissing is usually directed not against conspecifics but against unfamiliar adversaries, such as small predators that are nearly an even match in a confrontation. One fascinating encounter that I witnessed is unfortunately not recorded on film. It was a tame greylag goose and a small rufous-colored tomcat standing face-to-face and hissing at each other, in a perfect demonstration of universality of this display sound.

A hissing goose always orients its head so that it can fixate the object of the hissing with both eyes. It is noteworthy that geese also hiss at their offspring when they are engaged in rank-order fighting (Plate XIII-4). In our view, such hissing is elicited by the context "gosling attacked by a small predator," although the highly aroused adult geese must be having some sort of hallucination that a predator is present.

Reproductive Behavior Patterns

The behavioral system of reproduction includes many vocalizations and motor patterns that we have not yet mentioned. Some of the patterns are highly ritualized and occur only in the context of reproduction. Characteristically, they are immune to modification by conditioning and cannot occur in any stimulus situations other than the ones for which they have been phylogenetically programmed. In this sense, they count among the simplest instinctive behavior patterns. Many of the motor patterns, which are observed only during mating or nest building, can potentially be performed by both sexes, since the program for their performance as laid down in the central nervous system is the same in both males and females. This is true of many vertebrates—for example, fish—and it also applies to those species in which sexual behavior shows a pronounced sexual dimorphism. Nest-building activities are observed in ganders the same as in female geese, though at a lower intensity. Further, males are sometimes seen making a nest hollow, although they rarely are seen transferring nest material. As a general rule, however, it is the female goose that builds the nest.

Seeking a Nest Site

A firmly mated goose pair begins to show an interest in nest sites in early spring. The pair separates from the flock and travels widely. The gander accompanies the female wherever she goes, but he is not directly involved in the choice of the nest site. Eventually, the female chooses a site, for reasons that remain obscure, and begins to build a nest. Sites that afford favorable conditions for certain instinctive behavior patterns, such as the transfer of nest material and the nest-hollowing pattern, exert a particular attraction. The nest must also be located close to water and should be partially shielded yet permit good all-round vision. These conditions are often found on islands with sparse, low vegetation.

Geese sometimes will do without the clear field of vision, however, and will establish their nests in special boxes we have constructed as islands in the water. They obviously have no understanding of the ease with which a fox or other predator can reach a nest. Many of their nests are far too accessible—located, for example, in natural hollows between tree roots in a wood, among reeds, or on exposed tussocks (Plates VI and VII).

The female goose often remains on the nest site for a

time before laying any eggs. According to Heinroth, she does this to check on the security of the site. It can nevertheless happen that a goose will take a site to be secure when that is so only at a certain time of day, as along a path unused in the morning but regularly used in the afternoon.

Often the pair maintains several potential nest sites at once. Because nest-building activities are largely independent of external circumstances, the location and the exact center of the nest may not be determined until the first egg is laid.

Transferring Nest Material

Young geese of both sexes are often seen carrying straws or other pieces of plant material in their beaks and passing them back over their shoulders (Plate VIII-2). Even geese that are not involved in nest building can be observed upending to gather underwater plant material and then passing it back over their shoulders. In 1930, we often saw this with ganders who were performing neck dipping as a prelude to mating, but we have not seen such neck dipping followed by the transferring pattern since then.

The transferring pattern occurs more frequently as the female begins to show reproductive motivation in the early spring. As soon as the goose has selected a nest site and begun repeatedly performing nest-hollowing movements there (described in the next section), her orientation to the nest site assumes an important role. The goose shows the transferring pattern only when standing or lying with her head pointed away from the center of the nest. The effect of this orientation response is that all the nest material finds its way to the center of the nest.

Anatids do not have the capability of carrying material to the nest from a long distance away. When Heinroth once tried to facilitate the breeding of his geese by fixing them nest baskets in trees, they would perform the transferring pattern only with such objects as pieces of bark and small twigs that they could reach while standing on the nest site.

This limitation in anatids is all the more remarkable in that other bird species, no more highly developed in terms of the variety and adaptability of their motor patterns—and even certain fish species—show greater skill in nest building. A coot can place a suitable reed in its nest even when it requires a forward movement of the head. Similarly, the yellow-headed jawfish (*Opistognathus*), a relative of the gobies, builds its nest tube with perfect sense. It laboriously collects stones and places them regularly in the correct spots—where they are most needed. The large pile of plant material found in a swan's nest, however, is accumulated entirely by means of the transferring pattern. Although it may seem unbelievable when we examine a towering swan's nest, transfer of nest material and nest hollowing are the only nest-building patterns that anatids perform.

The Nest-Hollowing Pattern

In the nest-hollowing motor pattern, the bird raises itself on "all fours" by pressing backward with the feet and forward with the wrists—the wing shoulders—in an outward pushing motion (Plate IX-3). Simultaneously, considerable pressure is exerted between the wing shoulders as the bird thrusts its breast forward over the ground. With the aid of this movement, which is always performed after the bird has settled on the nest site, it gradually forms a hollow in the nest by pushing material from the center in five directions at once.

This basic nest-hollowing pattern seems to be common with all the birds I am familiar with, ranging from canaries to chickens and night herons. Many birds, including the greylag goose, invariably perform the pattern when they settle down to brood.

The Egg-Rolling Pattern

Like many other ground-nesting birds, the greylag goose is equipped with behavior patterns for returning an egg to the nest if it slips out of the hollow. While standing in or close to the nest, the goose moves slowly and carefully forward with the neck outstretched until the tip of the beak makes contact with the egg. Maintaining this contact, the beak is passed over to the far side of the egg, and the egg is rolled back into the nest hollow by pressure from both sides of the lower jaw (Plates IX-1, IX-2).

The movements performed by the neck and head in the egg-rolling pattern probably are innately coordinated, since they are constant both in their form and in the muscular effort exerted. By contrast, the sideways movements of the head that keep the egg balanced on the underside of the beak during the rolling process are demonstrably governed by tactile stimuli emanating from the egg. But the path the egg follows while being moved extends only from the point at which the underside of the beak first makes contact with it to the point at which the egg comes to rest against the feet of the standing goose. At this juncture, the goose seems reassured and returns to the center of the nest. If the egg is still visible on the edge of the nest or beyond, the process is repeated. In other words, the goose cannot balance the egg on the underside of the beak and walk backward until it has reached the nest hollow. Yet many ground-nesting birds can do just that. To elicit the egg-rolling movement, an object need merely have a flat surface with few irregularities. Large protrusions are immediately grasped in the goose's beak and nibbled. Unfortunately, a goose often shows this response toward an already pipped egg, which has sharp edges protruding from the opening in the shell. Vocalizations produced by the goslings in the process of hatching probably inhibit the egg-rolling pattern at this time.

In the experimental part of our study of the egg-rolling pattern, Niko Tinbergen was able to show that the direct backward movement that performs the work is based on central coordination, while the sideways balancing movements are orienting movements. The form of the direct backward movement is constant and is blocked whenever a modification is imposed. Like the form of the movement, the muscular effort exerted is also virtually constant. Objects somewhat lighter than a goose egg were lifted off the ground, while objects a little heavier served to block the movement, despite the fact that in other situations a goose's neck can exert a far greater muscular effort than egg rolling requires.

The orienting nature of the sideways movements is immediately obvious when one realizes that they are completely absent in egg rolling that occurs as a vacuum activity. They are also absent when a goose is presented with a wooden cube of about the same weight as an egg. Because it can be neatly balanced on the underside of the goose's beak, the cube does not deviate from the direct backward path.

The stimulus that elicits the orienting balancing movements can also be controlled by using "rails" to fix the egg's path. We once set out two bundles of reeds positioned so that they led from the egg obliquely past the nest instead of into the nest hollow. The goose rolled the egg along this path only until it had reached the point closest to the center of the nest. Then, the beak lost contact with the egg and was brought back to the nest hollow in a vacuum response.

The mechanism that triggers the egg-rolling response is relatively unselective. The appetitive behavior that precedes it—the oriented extension of the neck—is elicited by any object with a continuous and reasonably smooth surface. The size of the object can vary, within wide limits. A goose would initially accept an Easter egg made of papier-mâché and 25 centimeters long, although the response was extinguished after a few attempts that led to blocking, which is apparently very frustrating. It is thus clear that the releasing mechanism can be refined through learning.

Nest Lining, Covering the Clutch, and Egg Turning

During incubation, the clutch of eggs usually rests on a layer of down feathers, making one think that a solicitous mother has been at work (Plate VIII-1). It is sometimes said that the mother plucks the down feathers out of her own belly, but we have never observed such behavior. Rather, we commonly see down feathers fall onto the nest when the goose returns to the almost completed clutch after bathing. We have never seen a goose tucking down feathers into the nest after she has returned to it.

On the other hand, we have repeatedly observed the goose carefully covering the clutch with nest material before taking a brooding pause of any length. This is done in such a way that the down feathers on the inner wall of the nest are placed over the eggs first, followed by tiny twigs and leaves, which provide both insulation and camouflage (Plate X-1). This behavior pattern clearly involves visual monitoring and continues as long as any portion of the light-colored eggshells or down feathers is still visible. Even the very first egg is carefully covered in this way before the goose leaves the nest.

When the goose returns from a brooding pause, she seems not to know exactly where to find the center of the carefully hidden nest. Within a few paces of the nest, she begins to place her feet very carefully, as if "walking on eggs," and continues so until she is actually standing on her clutch. The eggs are located with a small turning motion of the beak, and then the goose settles down with a cautious shuffling movement. This is followed by a number of particularly intensive nest-hollowing movements, often accompanied by egg turning (Plate IX-4). Even the eggs of an incomplete clutch are turned whenever the goose returns to the nest, and later they are turned several times a day. Once the goose is firmly installed on the nest after a brooding pause, she often tucks nest material under her body by the use of her closed beak (Plate X-2).

Nest-calling

Nest-calling is a composite vocalization that incorporates a number of readily recognizable elements, including distance calls, alarm calls, departure calls, and rolling calls. The elements have become so fused through ritualization that nest-calling is immediately recognizable to the human ear as the same vocalization, no matter what the circumstances. The famous sentinel behavior of domestic geese is in fact entirely due to their readiness to engage in nest-calling. In some geese, this vocalization is clearly dominant over all others and is elicited by a wide variety of stimuli, although mild disturbance is the most effective cause. It was unquestionably the nest-calling of the geese of Rome that warned of the enemy attack.

As a characteristic outcome of extensive ritualization, nest-calling has become dramatically exaggerated in another goose species, the domesticated swan geese (*Anser cygnoides*). These intensively selected geese, which Heinroth has described as "falling over backward with their arrogant bearing," perform nest-calling with such persistence that they easily get on one's nerves.

Nest-calling is quite variable, and for this reason recognition of individuals is easy. Its primary function is to summon the mate when a brooding goose is disturbed (Plate IV-3). If he seriously intends to defend his mate, the gander responds with the distance call and flies quickly to the nest, usually while uttering rolling calls. A precise analysis of the components integrated into the female's nest-calling would provide a reliable indication of her general motivational state.

Incubation

After laying her last egg, the female goose begins to incubate. The gander keeps a short distance away from the nest and flies up to it only if the goose utters nest calls. The weaker the adversary, the more vigorously he defends his mate. Northern geese, which are threatened mainly by other birds and by the Arctic fox (*Alopex* sp.), are far more active in defense than are greylag geese, which are not very effective against the large red fox (*Vulpes* sp.).

A goose is not always incubating when she is sitting on

the eggs. During the egg-laying period, she sometimes lies on the nest with her belly feathers flattened, leaving the eggs unwarmed. In precocial bird species that lay several eggs, it is important that the eggs hatch in synchrony, so that all the offspring will be ready to leave the nest at the same time. Eckhard Hess demonstrated with the mallard that acoustic signals from the hatching ducklings pass from egg to egg and ensure this synchronization. It seems likely that a similar process takes place with the greylag goose.

During the incubation period proper, the goose remains almost motionless on the nest, often in the sleeping posture. From time to time, she interrupts her rest to turn the eggs, settling down afterward with extensive nest-hollowing movements. In particular with successfully hatching clutches, we have noticed that the eggs are oriented with the cracked area pointing upward. This indicates that the egg turning must cease at the beginning of the hatching process, although we do not know what stimuli are responsible.

During brooding pauses, the goose exhibits a characteristically hurried behavior, grazing with peculiar rapid plucking movements. Sometimes she moves rapidly across barren patches, performing vacuum plucking movements as she goes. Her movements during bathing and preening, before the return to the nest, display the same nervousness. After every brooding pause, the goose consolidates her nest with the so-called stuffing pattern, in which she presses the edge of the nest underneath her once she has settled down.

These essential interruptions of incubation vary in length from individual to individual, and the location of the nest site and its distance from the grazing grounds also influence the duration of the brooding pauses. We found that the geese we bred in nesting boxes in the middle of the lake might not leave the nest at all on some days. The constraints on her activities cause a brooding goose to lose weight markedly, and her feet become pale. These signs immediately betray any goose who has been incubating unbeknown to us and has abandoned her brood. It is my impression that geese are sometimes obliged to give up brooding because of their worsening condition. As the incubation phase progresses, the brooding pauses become both shorter and less frequent, and the goose does not leave the nest at all in the last twenty-four hours before hatching.

A goose's behavior toward any human who disturbs her during brooding depends on how tame she is. Very shy geese flee without uttering a nest call. Tame geese that have no specific relationship with the person who disturbs them will flee while nest-calling loudly. Very tame geese that are bonded to a human caretaker will respond in a friendly fashion. One particularly tame goose once raised her wing to let me place a gosling beneath it. Very tame geese that do not have a particularly close bond with the person who disturbs them will defend the nest with their wings outstretched, hissing and sometimes striking out with their wing shoulders.

A gander's behavior in response to human disturbance also depends on the degree of tameness. Only the tamest ganders will actually attack someone near the nest, and even this is rare. Most ganders confine themselves to approaching and loudly participating in the nest-calling of the female.

When a goose takes a breeding pause, the gander usually joins her, performs the triumph ceremony with her, and accompanies her back to the vicinity of the nest at the end of the break. One gander named Benjamin who was mated with two look-alike sisters (Verena and Röschen) habitually led either of his mates back toward Verena's nest. If the companion happened to be Röschen, he would attack her and chase her away at the last moment. This led me to believe that he was having as much difficulty telling the sisters apart as I was.

The gander never comes up to the nest during the incubation phase, but he maintains contact with his incubating mate by visual and vocal means. In some way he gets to know about the hatching of the goslings and appears at the edge of the nest as soon as the first little heads have emerged from beneath the mother's body. We do not know whether he hears the goslings or whether the mother communicates with him in some way.

Inhibition of Rapid Walking

One characteristic that the greylag goose shows on leaving the nest, which is very important for the cohesion of the flock of goslings, is that her locomotion is markedly slowed. One domestic goose with a bad limp turned out to be an especially good mother for this reason. In the gander, slow walking is an ability that can be reinforced through learning. We had one goose that formed a trio with two ganders bonded to each other. She led her offspring reasonably well, but she always preferred to stay behind the two ganders, and this had an unsettling effect on the family. Only after the two ganders adapted to slow walking did the difficulties disappear.

Sheltering the Offspring

While the goslings are small, their heat production is inadequate to keep their body temperatures constant, and they must be warmed from time to time (Plate X-3). Different birds adopt different postures for this, and a small bird with many offspring will literally be pushed upward by its brood. If one can look at a mother partridge from the side, one will see a forest of little legs bearing the bulge of the mother's belly. A mother goose sheltering half-grown offspring often is unable to lie down properly and is stuck with her belly somewhat raised.

The extent to which a goose raises her wings while sheltering offspring depends on the strength of her general drive to perform sheltering behavior. One very tame goose provided evidence that wing raising is an active process. If I held a gosling in my hand and moved it toward her from one side, she would raise the wing on that side. My observations indicate that the goslings, for their part, are slow to learn that it is easier to move under the mother from behind, against the direction of the feathers, than from the front of the wing. Experienced, somewhat older goslings will often purposefully lift up the trailing edge of the mother's wing.

Sheltering behavior apparently is strongly influenced by the motor display patterns and vocalizations of the goslings. Only on this basis can one explain why sheltering occurs much less frequently in warm weather than when it is cold or rainy. Dozing calls and, in more pressing cases, distress calls elicit sheltering behavior from the mother. As a rule, the mother's readiness to shelter the goslings disappears when they are about three weeks old, although in bad weather, with temperatures down around forty degrees, four-week-old goslings will still be allowed to creep under the mother's belly. Later, the mother goose refuses to shelter by simply standing up, and the goslings form so-called sleeping heaps.

The father is usually not involved in warming the offspring. One exceptional case is therefore of special interest in seeming to confirm that learning plays a considerable part in sheltering behavior. In 1982, the mother of one goose family (the "wild ones") was snatched by a fox immediately after her goslings had hatched. The father remained close to the goslings and protected them very effectively against attacks by other geese. At first when the goslings tried to creep beneath his belly, he simply lay down and the goslings pressed against him. Interestingly, in their attempts to get underneath him, they moved against the direction of his feathers. If they happened to slip under one wing, he would lift it up and they would move underneath. In this way, the gander gradually learned the complete sheltering pattern. But the most remarkable feat was still to come. By the next breeding season, the "wild one" had found a new mate. Despite the fact that this mate showed perfect sheltering behavior, the gander also performed sheltering. It is extremely unusual to see two goose parents sheltering their offspring side by side.

Neck Arching, Neck Dipping, and the Frigate Posture

While the bent-neck posture represents an invitation to form a lasting pair bond, the behavior patterns to be described now are exclusively concerned with invitations to mate and have no further significance for cohesion between the partners. The neck-arching pattern is a demonstrative display that is frequently performed by immature geese and is therefore not easy to interpret. It undoubtedly indicates a readiness to mate, but it is also a demonstration of masculine prowess. And yet I have never observed a releasing function of this behavior in the sense of a challenge to fight. Even if a female responds to this intensive pattern, which sometimes happens, it is no indication of a subsequent pair formation between the two geese.

The neck of the swimming male is curved into an elegant arch, with the feathers oriented in a special way that makes their edges strongly emphasize the shape of the neck. It is our belief that the ritualized pattern of neck arching is derived from the binocular gazing at the bottom of the pond that precedes upending. If the goose is strongly motivated for mating, the neck-arching pattern is combined with neck dipping. Neck dipping is undoubtedly derived from upending, and we can observe every conceivable intermediate between the two patterns. This accounts for the great difficulties in motivational analysis that we mentioned earlier. The motor patterns of neck arching and neck dipping are also shown by the female, though to a lesser degree. As a rule, the partners perform both patterns while swimming.

With increasing arousal, neck dipping and neck arching become more intense. The movements are performed ever more quickly, and sometimes the head is turned sideways underwater, as in bathing, so that a gush of water is thrown backward over the goose's body as it resurfaces. But we also know of cases in which the intense neck dipping that leads to copulation has given way to the transferring pattern associated with nest building, with the bird using small items picked up from the bottom.

At the beginning of neck dipping, the gander generally adopts a special demonstrative posture. He swims with his body, especially the rear and the tail, raised peculiarly high on the water. It is reminiscent of the shape of old sailing ships, and we have therefore coined the term frigate posture to describe it. In this posture, the wings are somewhat raised from the shoulder joints and slightly spread (Figures 66, 67, and 68).

Mating and the Postcopulatory Display

After neck dipping has reached its maximum intensity, the partners arrange themselves so that the female is oriented at a right angle in front of the male. She gradually flattens herself out on the water, inclining slightly toward the male. He then mounts her and performs copulation, while grasping her neck feathers in his beak (Figures 69, 70, and 71).

Afterward, the male slips backward into the water, and the partners then turn toward each other and raise their heads, necks, and folded wings high in the air (Figures 72 and 73). The intensity of this postcopulatory display decreases with the degree of familiarity between the partners. The display almost disappears in long-established goose pairs. The female is the first to stop showing the display and begins to bathe shortly after copulation.

The frigate posture, neck arching, and neck dipping are all seen frequently in geese that are only a year old, and copulation also occurs on occasion, although full sexual maturity is not reached until the age of about two years. In contrast to the first tentative signs of triumph-calling, these copulatory relationships are no indication of future pair formation.

Figures 66, 67, and 68: *The frigate posture, neck arching, and neck dipping. At a high level of arousal, copulation ensues.*

190

Figures 69, 70, and 71: *Mating* . . .

Figures 72 and 73: *. . . and the postcopulatory display.*

Ethogram 2

We turn now to those behavioral systems marked by social feedback between conspecifics. The society of the greylag goose has an extremely complex structure, which is reflected in the behavioral interactions between all the members of the group. The contact call in all its different forms, through interactions among all the personalities of the group, generates a multitude of social functions that make up this structure.

The Contact Call and Bonding

Helga Mamblona-Fischer refers to a certain call containing two or more syllables of variable sound intensity as the *vee* call. "This call is uttered when an egg is warmed after cooling, when a gosling is breaking through the eggshell during hatching, when the dry egg membrane is moistened, and above all as a response to noises." Because the pitch of the human voice is quite close to that of goose vocalizations, human speech readily elicits the *vee* call, even from goslings still in the egg. "Within certain limits, the louder the speech, the louder is the response and the greater the number of syllables (up to a maximum of four). When several goose eggs have been chipped in an incubator and a *vee* call comes from one of them, goslings in the other eggs respond with the same call. The greater the number of *vee* calls produced, the louder are the responses."

The *vee* call later develops into the contact call, represented by various forms of cackling. It should be noted that the contact call initially consists of two syllables but is soon produced in more. The infantile stage of cackling can be elicited immediately after hatching by any sizable sound-producing object. After a few days—usually the third day after the goslings leave the nest—the contact call can be elicited only by the parents and shortly thereafter by the individually recognized siblings. Selectivity in the release of this call has been progressively refined to restrict it first to conspecifics and ultimately to a small number of individuals.

It is a matter of definition whether we interpret the contact call as an expression of bonding to particular conspecifics or as an instinctive motor pattern with a strong appetitive drive that leads the goose to a specific consummatory situation that is communication with a partner. But one can hardly exaggerate the importance of bonding through shared contact-calling or cackling. With reference to the great apes, Robert Yerkes once said: "*One* chimpanzee is no chimpanzee at all." The principle, which also applies to the human species, is equally applicable to geese. A goose that is deprived of any opportunity to communicate with conspecifics is a pitiable cripple condemned to silence.

The Role of Personality

The remarkable increase in selectivity of the contact call, which occurs when the instinctive motor pattern is initially evoked by almost any creature but is later confined to one or a few conspecifics, renders any given individual irreplaceable in its social relationships.

The use of the term *personal* in humanities studies is often criticized. Originally, the word *persona* was used, in the sense of "mask," a reference to the roles played by the actors in classical Greek drama. But such a distribution of roles is in fact expressed in greylag goose society, and in an exemplary fashion. The essence of personality is surely present when the role played by one individual in the pattern of interaction between conspecifics cannot just be taken over by another. The defining property of a person doubtless resides in the fact that substitution is impossible.

The degree of bonding that is produced between two individual geese by their participation in shared contact-calling can vary considerably. As a consequence of certain ceremonies that we will describe shortly, the partners of a pair are bound more strongly than siblings are. Nevertheless, siblings frequently stay together, on the ground and in the air, for years after fledging.

The strength of the "rubber band" that pulls two individuals together might be measured in terms of the distance they customarily maintain between each other. For various living organisms, "walking out together" is often the first sign of a developing relationship. Helge Böttger has shown that the distance between two individuals while resting is a useful indicator of their degree of bonding. Indeed, we might conclude that there is a balance between attracting and repelling forces when two animals sit or lie down, for it is impossible to discuss the attracting force involved in bonding without also considering the repelling effect of aggression.

Aggression

A peculiar relationship exists between bonding and aggression. It apparently has survival value for individuals of a species to repel each other, in that they become distributed uniformly over the available habitat. Aggression, along with the territoriality it engenders, is one of the most important mechanisms for the dispersal of living organisms. We know of many animal species in which the individuals only repel each other and do not show bonding of any kind. But we know of no animal species that practices bonding but completely lacks the aggression required for dispersal.

The processes involved in individual recognition seem to be subject to particularly strong selection pressure when two conspecifics cooperate in the care of their offspring. Under these conditions, it is advantageous for the species if aggression between two individuals cooperating as parents is completely eliminated but, on the other hand, maintained or even enhanced against other individuals. That has been shown to be the case with pairing fish, in which the mechanisms of aggression and its deactivation between partners are well understood. In the fish studied by Rosl Kirchshofer, individual recognition leads to a phenomenon of reciprocal attraction. When all the individuals repel each other, a lesser degree of repulsion has the same effect as attraction. The opposing forces of attraction and repulsion—the "biphasic processes underlying approach and withdrawal" of Theodore C. Schneirla—are here expressed to the full.

Accordingly, any two greylag geese will be engaged in a conflict between attraction and repulsion. The former is minimal between adult geese that are complete strangers to each other. Nevertheless, if two such geese have no other possibilities available, they will show a weak cohesion. The strongest bonding occurs between family members, and particularly between the partners of a pair, as is generally the case in individuals that have developed a joint triumph ceremony.

Motor Patterns in Aggression Between Rivals

Forward stretching of the neck (Figure 74) seems to be the phylogenetically primitive expression of aggression in anatids. In some species, for example the shelduck (*Tadorna* sp.), even freshly hatched ducklings show forward stretching of the neck in threat during conflicts with siblings. The persistent conflicts

Figure 74: *Direct forward stretching of the neck: pure aggression without fear.*

between goose families make it nearly always possible to see many individuals in a large flock with their necks stretched forward. Ganders leading families are often seen making room for their dependents by pressing forward with their necks outstretched. But the frequency of threat behavior can easily be overestimated because, as we will see, a slightly reoriented forward-stretching of the neck is also employed as a gesture of peaceable greeting.

In addition, the posture of the outstretched neck is modified when pure threat is supplemented by other kinds of motivation, particularly the tendency to flee. If an advancing gander meets resistance from an opponent of roughly matching strength, his partial motivation to flee will be apparent from the posture of his neck (Figure 75). The neck is compressed backward to form an upward arch in the sagittal plane, while the head remains horizontally directed toward the opponent. As has already been discussed (page 94, and as illustrated schematically in Figure 7), it is possible to recognize

Figure 75: *Threatening posture of the neck with a slight upward bend reflects the mild influence of the escape motivation.*

a series of characteristic neck postures that reflect the spectrum of motivations for attack and flight.

In extreme cases, backward compression of the neck can modify the neck's threat posture to produce a peculiar orientation that we originally labeled the elephant neck. (This somewhat misleading term alludes to the shape of the elephant's trunk when it is hanging down.) When showing this pattern, the goose stands with its legs very straight, its wings stretched, and its neck held almost vertically downward, but with the head pointing toward the opponent (Figure 76). This posture indicates that the gander will advance no further but also is not going to retreat and intends to defend his position with blows from the wing shoulders.

When two ganders are so evenly balanced in strength and motivation that neither will give way, a wing-shoulder battle ensues. The opponents face each other, and each grabs the other with his beak, usually by the shoulder, trying to force him backward and downward. The gander rears upward, regardless of whether the fight is taking place in the water or on land, and the next instant strikes out at his opponent with

Figure 76: *The elephant-neck posture in the bean goose.*

one wing. The wing is raised from the body, but is kept half bent at the elbow joint and fully bent at the wrist joint so that the horny fighting spur sticks out prominently. The other wing is spread fully and is stretched asymmetrically straight backward. In effect, one wing serves as a sort of fist and the other as a counterpoise (Plates XI and XII). We often under-estimate the strength of these blows. On one occasion, in the Berlin Zoo, a blow of this kind from a mute swan broke both the forearm bones of a keeper.

Ganders can strike out most effectively with their wing shoulders while in the air. I cannot say whether the rivals grasp each other with their beaks, for the process takes place too fast. I was once subjected to an aerial attack from a Canada goose (*Branta canadensis*) in which I received a literally stunning blow on the head. I can only record that the gander flew off after the attack without touching the ground. I do not know whether he grasped my hair in his beak. In the course of one fight between two greylag ganders, one of the combatants was struck so violently on the side of the neck that the nerve plexus of the wing (*plexus brachialis*) was completely paralyzed. He plummeted at least ten meters straight down without

Figure 77: *The extreme lowering of the neck shows a strong inhibition through the escape motivation.*

making the slightest wing movement, but luckily he landed in deep water and managed to climb out, his wings drooping.

Nervous coordination of the striking movement with the wing shoulder is developed in young goslings only a few days after hatching. It is touching to watch as one of these balls of down stretches one wing backward as a counterpoise and performs stabbing movements with the other, which is bent at the wrist joint. And consider that at this stage of development the wing does not even reach the middle of the gosling's chest and is certainly too short to reach an opponent (Plates XIII-1, 2, and 3).

When there is conflict between the motivations for attack and escape, a different threat posture may be displayed, leading eventually to a wing-shoulder battle. In this case, the neck is stretched forward somewhat further and the nape lowered more markedly, so that the neck is held low and parallel to the ground (Figure 77). In this posture, the direction of the threat is oriented away from the opponent, with the result that the two opponents can approach each other in a broadside orientation (Figure 78) until they are

Figure 78: *Broadside threat with the neck stretched forward and close to the ground.*

standing parallel. An exchange of blows from the wing shoulders, accompanied by grasping with the beak, follows abruptly. I have observed this threatening approach only between high-ranking ganders, who rarely fight each other. In this situation, it may also happen that one of the opponents loses courage and flees by jumping over his adversary.

The same motor patterns can be observed in greylag ganders that are leading young goslings, though actual fighting is exceedingly rare. In June 1986, I observed two ganders approaching each other in the broadside orientation just described. I had never seen a fight between these two high-ranking ganders, and therefore I was extremely interested. The two drew ever closer together, stretching their necks further and further and lowering them toward the ground, and their eyes looked as though they were about to pop out of their sockets. I was expecting a vicious fight to break out at any moment, when suddenly the ganders froze on the spot as if turned to stone. This cramplike standoff eventually came

Figure 79: *The goose on the right retracts its head in surprise.*

to an end, and the ganders moved silently apart. Shortly afterward, I saw the mates of the two ganders grazing peacefully with their flocks of goslings only a few meters apart.

One rare, low-intensity form of threat occurs when a gander suddenly becomes afraid of another, particularly after having just performed a demonstrative display. The resulting head retraction, performed so to speak "in surprise," is illustrated in Figure 79.

Whenever neck stretching is elicited in the presence of a conflicting escape motivation, resulting in the retraction of the neck, there is a lateral trembling of the neck. The movement is most pronounced at the point where the retracted neck is bent, and it always results from a conflict between two kinds of motivation that is taking place on the periphery of the central nervous system. Neck trembling is often seen when a tame goose is trying to take food from a human caretaker's hand but does not dare stretch its neck far forward. We also see it when a goose stretches out its neck to bite an opponent it is somewhat afraid of. Finally, neck trembling can occur in situations in which there is no intention to grasp anything with the beak, as when a goose stares with both eyes at an obstacle before climbing or jumping over it. Whenever neck trembling occurs, it is possible to recognize the strong arousal of the sympathetic nervous system from the goose's bulging eyes and flattened feathers.

We refer to one peculiar motor sequence, which is always preceded by neck trembling, as "jabbing over." The goose silently approaches a motionless conspecific or some other creature, fixating it with both eyes. It then pulls its head far back and, in a lightning fast movement, jabs its head and upper neck forward with the beak closed, passing just over the top of the object. The movement is carefully aimed but nevertheless clearly inhibited. I originally believed that geese performed this movement only with sleeping conspecifics, but this has proved to be incorrect. One possible interpretation is that a goose showing this behavior is attempting to induce a sleeping conspecific to withdraw its head from its shoulder feathers and permit itself to be recognized. Young goslings perform jabbing over only with siblings, and it seems that a certain degree of familiarity may be necessary for the response to be elicited in older geese. The movement is usually performed over the back feathers of a conspecific, but it is sometimes directed over parts of the human body, such as the arm, foot, or back of a caretaker lying in the grass. We have never seen jabbing over performed with inanimate objects. It occurs only with objects that are not moving but are known to be both mobile and alive. The motor pattern is apparently very ancient in phylogenetic terms, as it is common to all geese and ducks.

Submissive Behavior

In many animals, behavior patterns that minimize the likelihood of an individual's eliciting an attack are, in a manner of speaking, probably negatively selected in relation to the func-

Figure 80: *The sneaky posture represents a complete absence of attack motivation.*

tioning of fight-eliciting mechanisms. In fish, a broadside presentation with the dorsal fin raised while the fish swims in a jerky fashion represents the strongest fighting challenge. The most submissive behavior shows a converse pattern. The fish makes itself as small and slender as possible, by lying on its side and folding its fins, and moves slowly and stealthily.

In greylag geese, the forward stretching of the neck in threat represents a challenge directed at the opponent. In the submissive posture, conversely, the neck is retracted as far as possible, so that the head comes to rest on the bird's back (Figure 80). We have come to use the term sneaky posture for this pattern. Intimidated geese, especially those that have lost a mate, sometimes show the sneaky posture for months on end. The submissive posture can be exaggerated if a threatening conspecific comes close. The same general body posture is maintained, but the back of the head is turned toward the opponent. The sneaky posture is generally characteristic of individuals that are low in the hierarchy and unwilling to fight.

An unusual posture of the neck and body that we are

Figure 81: *Swollen-neck posture. The bird on the left will not attack but is also unable to flee.*

inclined to interpret as submissive is seen when a gander is attacked but does not want to fight. The gander raises its neck high in the air and ruffles its neck feathers (Figure 81). We have seen this posture especially at times when a gander has been attacked by a rival but is unable to escape, perhaps because it is close to its nest. Our provisional interpretation is that this swollen-neck posture reflects an unwillingness either to fight or to flee: "I would like to leave, but I can't."

Figure 82 shows the posture of total submission after defeat in a drawn-out fight.

The Social Hierarchy

One mechanism that is particularly important for reducing aggression is rank order. Many years ago, T. Schjelderup-

Figure 82: *The posture of total submission.*

Ebbe discovered the so-called pecking order of domestic chickens: decisive fights between individuals take place only once or twice, and thereafter the loser gives way to the victor without opposition. As a general rule, the stronger individuals are dominant and the weaker individuals subordinate, although circular relationships can exist. The relationships often persist for years, indicating that an individual, once it has been defeated, accepts its subordinate position and makes no attempt to change it. There is no question but that the resulting rank order deactivates aggression. This mechanism, which appears to have survival value, is found in almost all higher animals, including crustaceans, insects, fish, birds, and mammals. Characteristically, the rank order, once established, is more stable in animals with simple behavioral capacities than in animals with complex central nervous systems. In fish, a physically stronger individual may remain subordinate for years; in wild boar, a few days of separation are sufficient for a defeated individual to return with new hope of success.

Coexistence in a hierarchically organized society leads to familiarity among the members, and this raises the threshold for the elicitation of aggression among the members. Put another way, the members of a society will attack strangers more vigorously than they will individuals with whom they have become familiar over a long period, regardless of whether the latter are dominants or subordinates. This can even lead to a situation in which the members of a group will support each other in a fight with strangers. In ponds at the oasis of Gafsah, males of the cichlid fish *Haplochromis desfontainesii* (whose males are attached to fixed sites) construct their spawning trenches close together. The effect is that every territorial male becomes so accustomed to his nearest neighbors that he attacks them far less vigorously than he does strangers. As R. Kirchshofer has demonstrated, resident males attack strangers very aggressively, sometimes joining forces in the attack. The behavior is actually a very appropriate one, since a familiar neighbor already has his spawning trench and therefore poses no territorial threat.

Rank Between Families

The rank order within tightly cohesive groups of greylag geese is complicated by the fact that members of groups or families will support each other. In a conflict between two geese, the decision whether to attack or flee may therefore depend on the presence of group members and on the number willing to join in to support either antagonist. Many conflicts between families are sparked by a gosling that dares to threaten another family. The gosling is threatened in turn, and its family members then enter the fray (Figures 84 and 85). When the rank between families of greylag geese is

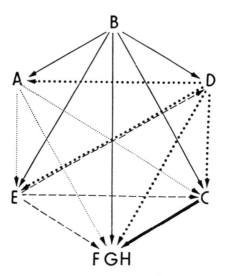

Figure 83: *Rank relationships between eight families. The directions of the arrows indicate which family is dominant (that is, → indicates that B is dominant to A).*

determined in the rearing area, we find a relatively constant ratio between attack and escape among the dominant and subordinate individuals (Figure 83).

To quote Angelika Tipler-Schlager's explanation of this diagram: "The arrows represent the directions of attack and flight among the combatants. The diagram also takes into account observable criteria derived from our ethogram (aggressive behavior, active and passive avoidance). If a given group proves to be dominant for a period of time, this is regarded as indicating a clear rank relationship. It is the ratio between aggressive behavior and escape responses that establishes rank. High-ranking geese flee far less often than low-ranking geese or geese whose rank is unclear. Aggression is not correlated with dominance. The rank of a family does not depend on the age of the offspring, as was assumed by Heinroth in 1924. Also, the number and sex of the goslings have no influence. Finally, the sequence of arrival [at the rearing area] has no influence on the rank of a family."

Figures 84 and 85: *One gosling attacks a gosling from a different family, chases it away, and returns to its own family with a greeting.*

Heinroth noted that pairs with young offspring are usually dominant on their particular ponds, but this may be an oversimplification. Additional, long-term observation is needed. For one thing, one should determine the influence of the former rank of the parents on the relationships between the families.

Figure 86: *A wing-shoulder battle between two ganders. Their families participate in the fight with roll-cackling, but they rarely become actively involved.*

Ultimately—one might say as a last resort—the rank relationship between two families is decided by a wing-shoulder battle between the respective ganders that lead them. In such a fight, the members of the entire flock gather to watch. The longer the fight lasts, the greater the number of onlookers (Figure 86). Interestingly, it is rare that one of them joins in. We have recorded only a few cases in which a male individual paired with a gander intervened to support him, and only two cases in which the female partner supported her mate.

Rank Within Families

Heinroth believed that complete peace reigned within the goose family and it was therefore inappropriate to refer to a family rank order. In 1951, I myself wrote: "In anatids, and particularly in geese, rank-free familiarity [among the goslings] is maintained until late autumn and only then does it give way to a rank order."

The reason skilled observers failed to see the development of a rank order among goslings is probably due to its surprisingly early beginning. Sybille Kalas-Schäfer was the first to discover it, in the early summer of 1971. She observed that although goslings form a tight anonymous association immediately after hatching, they begin to engage in fierce competition with each other after only six to eight days. The fighting usually occurs when the goslings awaken from a long rest—almost as if their indifference to each other while resting has extinguished their previous familiarity. With goslings that are reared by their parents, the fighting often occurs at night or in half-light and passes unobserved. Goslings that are led by a human foster parent usually fight at dawn or dusk, and Kalas-Schäfer was able to observe them by the light of a shaded heating lamp.

These first fights among goslings are informative with respect to the physiology and development of the motor patterns involved, because of the goslings' detailed matching of their patterns of coordination with those of the wing-shoulder battles of adult geese. Most obvious are the backward stretching of the tiny wing in a balancing movement and the flexion of the striking wing at the wrist joint, at the point where the horny fighting spur will develop some months later (see Plates XIII-1, 2, and 3).

The response of the parents to the fighting offspring is remarkable. Although they observe the proceedings closely with both eyes, they do not intervene. It is our belief that the parents anticipate, and may even be seeing as a sort of hallucination, a small predator attacking their goslings. In support of this hypothesis is the fact that the parents often open their beaks and hiss while watching (see Plate XIII-4).

The outcomes of these early fights to a large extent determine the rank order among the siblings. There is an accompanying change in the orientation of the neck. Prior to the rank fighting, the neck is always directed straight ahead toward a sibling; afterward, the newly established hierarchy is reflected by the direction of neck stretching. The gosling utters the contact call intensively and stretches its head obliquely past its sibling rather than directly toward it. The angle at which the head veers away corresponds to the degree of subordination the individual feels. Kalas-Schäfer has coined the term *evasive greeting* to describe the pattern. It can occur as a response to an aggressive behavior pattern, but it is also performed when no prior aggression has been directed toward the greeting gosling. Of course, the aggressive partner must also be a member of the family, an individual with whom contact calling is exchanged. As Kalas-Schäfer has established, the rank order determined by fights between the goslings during their first weeks of life persists until fledging.

When the geese are somewhat larger and their voices are beginning to break, a further indication of dominance and subordination emerges in the form of a varying intensity of greeting. In situations that elicit intensive greeting, the dominant greets far more vigorously than the subordinate. Conversely, when motivation is low in both geese, the subordinate gives the louder calls.

One exception to this rule is illustrated by the following incident, which I observed after the original caretaker of a flock of four goslings left and I took over their charge. The highest ranking of the four goslings was named Resi and the lowest ranking Mitzi. Shortly after fledging, these two geese emerged victorious from a fight with two young geese from another flock. Mitzi's victory was the more conspicuous because she had grasped the feathers of her opponent in her beak and been dragged across the pond, whereas Resi's opponent had abruptly dived away and left her. Accordingly, Mitzi was strongly motivated and came back across the pond toward Resi, greeting loudly. As is typical of geese after a fight, Resi started to bathe. But as Mitzi rushed up and loudly gave her triumph call in Resi's ears, the latter turned and snapped at her subordinate sibling, upon which Mitzi fell silent. An amusing scene followed: as Resi continued her bathing, Mitzi would give vent to triumphal greeting every time her sibling ducked her head into the water but would become modestly silent when the head reappeared.

The rank order that arises between goslings is not influenced by size or age, nor is it affected by the individual intensity of aggressive behavior. Jane Packard once sampled the behavior of four goslings she was rearing by counting the social patterns that occurred in the six possible paired combinations. In other words, she calculated how often a given gosling "said" something to another. A surprising number of individual differences emerged. For example, one gosling greeted significantly more often and more intensively than any of the others. Most important, though, was that the expected correlation between individual aggression and rank did not materialize. One female gosling exhibited a particularly high threshold for the elicitation of escape behavior, yet was rarely aggressive. Strikingly, this nonaggressive but fearless individual was relatively immune from attack by goslings of other flocks and tended to flee less often than more aggressive goslings did. Jane Packard's study produced many diagrams that constitute something of a guide to the character of greylag geese.

Just as aggression between rivals has no effect on their position in the rank order, a clear rank relationship provides no barrier to bonding behavior. Kalas-Schäfer has observed that siblings displaying a sharply defined rank order maintain tighter cohesion than do siblings that have been closely matched in fights and therefore have never developed a stable hierarchy. A certain degree of aggression persists among the latter siblings and prevents them from establishing close proximity. In Kalas-Schäfer's view, the stable rank order has survival value in that it reduces the social tensions to which individuals are exposed.

Different Forms of the Contact Call

The various forms of the contact call are independent of each other in that they characterize different forms of social interaction and also are acoustically distinct. On the other hand, the different forms of the call are linked by so many intermediates that it seems appropriate to group them under a single heading. All the variants of the contact call—even the rolling call—can be placed in the broadly defined category we refer to as cackling.

The contact call has two variable features. The intensity can change, rising from the faintest call to an intense greeting with neck stretching, and sometimes progressing to the triumph call. There also are qualitative changes, which lead from fluent cackling to roll cackling or group rolling, which seems to reflect an increase in aggression.

Greeting

When contact between two socially bonded geese is interrupted by some extraneous factor—they become separated or are distracted by an external stimulus that leads to vigilance—the contact call is temporarily silenced. After the disturbance is over, the geese resume their contact-calling at a somewhat higher intensity than before. Since this also happens when separated partners are reunited, we have come to refer to the behavior as greeting. Of course, that label is not appropriate in all cases. For example, geese also greet one another after they have displayed vigilance while a truck drove past.

When the contact call erupts again following temporary inhibition, the neck is stretched far forward and lowered closer to the ground. Contact calls accompanied by neck stretching cover a wide spectrum with respect to both the intensity of the motor pattern and the direction in which the neck is stretched. The latter, as we have said, depends on the rank relationship between the partners (Figures 87, 88, and 89).

Fluent Cackling and Pressed Cackling

Young goslings that are following their parents in a tightly knit flock produce a special form of the *vee* call. Sybille Kalas-Schäfer has chosen the term fluent cackling for this because the syllables follow in rapid succession and flow into each other.

Fluent cackling is the precursor of pressed cackling in individuals and is also a functional predecessor of roll cackling. It is characteristic of situations in which the parents are being greeted, but occurs as well during opposition to other families. Heinroth properly saw this as a distinct form of behavior and referred to it as family palaver, in which the offspring join in the cackling of the parents. "For the human observer," Heinroth said, "a family palaver of this kind appears extraordinary." Anyone who is at all familiar with the various aggressive displays of greylag geese will see that in this ceremony the offspring are supporting the cackling of their parents against another family. They direct their little necks toward the strange family—although usually while seeking shelter behind their parents. Nevertheless, a five-day-old gosling was once observed launching a genuine attack on a member of a rival family. An actual attack of this kind is always conducted in silence.

Pressed cackling retains the rapid sequence of syllables and the intensive neck stretching even after the voice has broken. It is also characteristic of the vocalization that its relatively soft sounds are forced out under high air pressure. But a strange phenomenon occurs when the offspring have grown to adulthood and, after the dissolution of the family, attempt to join their parents again—not at all uncommon,

Figures 87 and 88: *During greeting the neck is directed past the partner, with the deviation especially pronounced when the greeting is intense.*

Figure 89: *When the family engages in pressed cackling following a victory over another group, the heads of parents and offspring can almost touch if the greeting is intense.*

especially if the parents have been unsuccessful in breeding that year. To win the parents around, the offspring from the previous year utter fluent cackling at a surprisingly high pitch. Although they cannot quite match a gosling's voice, in a manner of speaking they are pretending to be youngsters again. Yet, a moment later, they can respond to other conspecifics in the deep tones appropriate to their age. We know of no other example in anatids of such a voluntary alteration of a display call.

With self-confident and aggressive hand-reared young geese, we have been able to observe at close quarters the transition from family palaver or fluent cackling to the rolling call. As the voice breaks, deeper and more raucous calls gradually become intermingled with the rapid tones of fluent cackling. The new elements are most likely derived from another independent vocalization, the rolling call.

Rolling with Cackling

In a certain sense, the rolling call is connected with cackling. Production of the rolling call doubtless depends on the presence of an individual that would also be a suitable partner for contact calling or cackling. In fact, the rolling call is performed only toward a conspecific that, in the words of Helga Mamblona-Fischer, "is involved in a relationship permitting the exchange of contact calls or is at least likely to be involved in such a relationship." She believes that the rolling call always reflects a conflict of motivation between bonding and aggression.

If, for example, one calls in the dark to a resting flock of geese on the other side of a large expanse of water—provided that a familiar individual is present—the geese respond with a series of extremely loud, brief trumpeting calls. I was initially inclined to believe that this vocalization, referred to in our institute jargon as the puzzle call, belonged to the category of alarm calls. But we later became aware that it was single-syllabled rolling calls followed by a number of additional truncated calls and then by genuine pressed cackling, just as in the triumph ceremony. When producing such calls, the partners of gander pairs turn their beaks toward each other and utter pressed cackling without any transitional calls.

A relatively pure form of the rolling call is heard and seen with young ganders that are just over a year old and are beginning to court the interest of females through demonstrative displays and attacks on specific opponents. The gander raises its head with the beak open wide and the hyoid bone depressed, making the neck look quite thick. The gander turns toward another male or, in the absence of a suitable partner, toward an imaginary adversary (see the description of Martin, page 25) and utters a "loud, cascading series of calls that are rich in overtones. Drawn-out calls, which are similar in sound and spectrographic appearance to alarm calls

Figure 90: *The sequence of movements in the classic triumph call.*

and distance calls but somewhat longer, are followed at
irregular intervals by short, truncated tones that resemble
cackling or contact calls." That is how Mablona-Fischer describes the introductory sequence of the triumph call. She
goes on to say that "the calling becomes more intense, in that
the tones become louder, longer, and more high-pitched, and
the neck is brought lower, as the gander rushes back to the
female after a foray against his opponent. While doing this,
he adopts a more exaggerated demonstrative posture: the
wings are spread wide, the chest is expanded, and the tail
feathers are spread. Sometimes he even flies back toward the
female. Just before reaching her, he slows the tempo; the
rolling call fades and gradually gives way to cackling, at which
point the female joins in" (Figure 90). The word "gradually"
in this quotation needs some amendment; between the last
rolling call and the beginning of cackling there is always a
brief but characteristic syncope.

The relationship between the rolling call and cackling is
complex. In the first place, the two kinds of vocalization can

be mixed. Despite the fact that they do not occur in a rhythmic sequence, it is my opinion that the truncated calls that are heard consistently at the end of a long rolling sequence derive their organization from cackling. Second, the roll cackling of a group of geese is most likely produced by the superimposition of pressed cackling on rolling. The pitch and intensity are derived from the latter, and the rapid sequence of syllables and the maximally stretched neck are derived from the former. Third, the sequence of rolling and cackling is ritually consolidated in one of the most important ceremonies in goose society, the classic triumph ceremony, which consists of aggressive rolling followed by cackling and accompanied by bonding motivation. Finally, I believe I have witnessed some transitional cases. These do not lead from rolling to cackling, but they do lead in the opposite direction. We can enrage a tame goose that has a triumph-calling relationship with a number of human observers and utters perfectly normal pressed cackling toward them by holding out a piece of bread higher and higher, making it more and more difficult to reach and finally impossible. The goose's unmistakable rolling calls become more and more mixed with cackling, and at the same time the goose starts to bite more firmly. This kind of personal rage toward a human is especially interesting in that normal geese with no bonding motivation utter the departure call when begging for food.

The Triumph Ceremony and Its Bonding Effect

The most impressive vocal display of greylag geese is the triumph call, which also is the most important factor in binding individual pairs together within the flock and hence in determining the flock's social structure. The rigid coupling of rolling and pressed cackling that has arisen through ritualization is obviously important for the bonding between the partners of any pair that performs the triumph ceremony. However much the process of pair formation may differ with individual geese, concerted triumph-calling is always the end result. When individual pairs are observed with their offspring—for example, Mercedes and Florian or Sinda and Blasius—we can see few differences in the behavioral interactions of the partners. The motivational force that drives each of these birds to perform the triumph ceremony with its partner appears to be overwhelming.

Heinroth has given this description of the triumph ceremony: "The simplest situation is one in which a pair of geese and their offspring approach a strange conspecific. The gander runs or swims angrily toward the strange bird with his neck stretched forward, and his adversary flees. The attacker turns around at once and hurries back to his mate [Plate XIV], at which point they both utter loud triumph calls. Blaring calls are followed by peculiar soft cackling that can be rendered as a continuous *gang-gang-gang-gang-gang* with a pronounced nasal quality. When performing this ceremony, the two geese look as if they are likely to attack each other at any instant, and they utter the loud calls directly into each other's ears, with their heads held just above the ground" (Figures 91 and 92).

The classic triumph ceremony the foregoing describes usually begins with a high-pitched, penetrating sound, which is immediately followed by rolling calls. After a brief interval, which is only rarely omitted, pressed cackling is uttered (Plate XV). One might speculate that the ritualized combination, or the cohesion, of rolling and cackling obliges the individual that has just uttered the rolling call with its partner to proceed to cackling and thus to complete the ceremony. We have seen, however, that the ritualized combination of rolling and cackling is not absolutely fixed.

Heinroth's description alludes to one point that emerged in our earlier discussion of the contact call. His phrase about "loud calls directly into each other's ears" indicates that the stretching of the neck in greeting is not directed at the partner as a threat but is obliquely angled away.

221

Figures 91 and 92: *A gander returning to his mate with classic triumph-calling.*

Roll-cackling or Group Rolling

One ritualized form of behavior that develops from family palaver and fluent cackling has been interpreted by Helga Mamblona-Fischer as a special form of the triumph ceremony. I prefer to treat it separately from the classic triumph cere-

222

Figure 93: *The ceremony of converging necks, or group rolling. The group is supporting the gander at the left, and he will respond by launching a renewed attack on a fleeing family.*

mony because it is a simultaneous performance rather than the ritualized succession of two elements. With the triumph ceremony proper, the attack is followed first by a discharge of the aggressive motivation through rolling calls and then, sharply separated by an interval, by an abrupt transition to pressed cackling, which presumably corresponds to the motivational shift from aggression to bonding. Pressed cackling is doubtless governed by a motivation that completely excludes aggression.

In roll-cackling, the same components are present as in classic triumph-calling, but, as noted, the two forms of motivation are expressed simultaneously. Whereas in the triumph ceremony the geese stand opposite each other with their heads crossed and call loudly "into each other's ears," in group rolling the heads of the geese are directed toward a rival group and oriented parallel to each other (Figure 93). This is why I referred to this ritualized behavior earlier as the ceremony of converging necks.

Figure 94: *Rolling with the chest expanded.*

. Group rolling occurs mainly in the autumn and winter, when many families or triumph-calling associations are combined into a flock for the autumn migration. Mamblona-Fischer gives this description: "The geese, compressed together in a small area, begin to utter rolling calls in response to the smallest thing that attracts their attention. They respond to conspecifics that are moving from one spot to another or merely changing their position, to any sudden noise, and the like. They usually adopt a demonstrative posture, inflating their chests, ruffling their head and tail feathers, and stretching their necks vertically or obliquely upward [Figure 94]. Rolling often begins when a single gander utters alarm calls. All the group members immediately join in with rolling calls. When this happens, entire groups turn toward each other with their necks stretched upward. Neighboring groups then begin to utter rolling calls as well, and in the densely packed flock the calling spreads from group to group like a chain reaction."

A goose under the influence of rolling motivation stretches

its neck vertically upward. The oblique neck posture described by Mamblona-Fischer arises from mixing in a threat posture directed toward an adversary and thus reflects a combination of rolling and threat. The larynx is markedly lowered so that the neck protrudes and appears thicker. The posture also causes two longitudinal lines to appear between the throat and the neck musculature.

Actual conflicts are common between adjacent groups during group rolling, although they are generally restricted to charges accompanied by biting-intention movements. Mamblona-Fischer has reported that other forms of the triumph call are rarely observed along with group rolling. When geese turn toward their own families in the course of the rolling-call ceremony, pressed cackling breaks through in their vocalizations, but the neck is not lowered to the horizontal position in the manner that is typical with this call. Further, the rolling calls do not disappear; rather, there is a superimposition of rolling and cackling. That is what gives the entire ceremony its characteristic quality. Following the transitory turning of the partners toward each other, the geese immediately turn back to face the rival group again. This makes the necks of the geese swing to and fro. These two directional tendencies often leave the members of a family or group standing side by side in a closed pack and pointing toward their opponents with their heads close together.

On one occasion, when several groups had returned together after a lengthy absence, many of the resident goose families repeatedly joined in a common front when performing the rolling ceremony. After the strange geese left, however, the resident group split into two opposing subgroups.

Roll-cackling actually represents a kind of war of nerves. Victory or defeat, advance or withdrawal, seem to depend on somewhat trivial factors. Many years ago, I recorded how the conflicts between two large and high-ranking families regularly resulted in victory for Kamillo's group over Schwestermann's. Yet the behavioral records of my colleague Heidi

Buhrow indicated the exact opposite. We almost began to mistrust each other until we discovered that the presence of one or the other observer was the decisive factor in the victory or defeat of a given family. When he was young, Kamillo had been a member of a flock that I was leading, which had also contained three snow geese, and now my presence greatly increased his self-confidence.

Under certain conditions, roll-cackling plays a decisive part in the determination of rank relationships between families. This is so when the geese have left their breeding grounds for the autumn migration. While our institute was located at Lake Ess in Upper Bavaria, we could attract the geese to pools in the nearby moorland by feeding them. They felt less secure there than "at home" and would not stay long without a familiar human companion to boost their self-confidence and without the lure of the food. The families remained cohesive and even before landing would utter numerous rolling calls along with their distance calls. A flock that had just landed would normally take up a convergent-neck formation as soon as it emerged from the water onto dry land, and its aggressive behavior was clearly apparent. Under these conditions, a correlation between the size of the family and the rank order was obvious. Also, neither I nor my coworkers can remember ever seeing a wing-shoulder battle. Unfortunately, in Grünau we are unable to lead the geese to a strange setting in which we might be able to observe similar behavior.

Rolling Without Cackling

Although we hear rolling in combination with cackling, in the classical triumph ceremony and in group rolling, we must regard rolling as an independent instinctive pattern, since it often occurs alone and in the absence of a ritualized context. There are many situations in which rolling can be observed without cackling, although we have been unable to demon-

Figure 95: *The rolling duet of a gander pair.*

strate whether the vocalization ever occurs in the absence of a potential partner.

In rare cases, rolling not followed by cackling occurs when a fight between the partners of a gander pair breaks out without warning. Gander pairs that remain continuously in close contact, reflecting a strong bonding drive, can produce drawn-out rolling duets from which all cackling is absent (Figure 95). An old, extremely aggressive gander that has no partner—possibly indicating an incompletely developed bonding drive—will persistently and exclusively utter rolling calls. Rolling in isolation, however, is far less frequent than the classical sequence of rolling followed by cackling in the triumph ceremony, and rolling doubtless serves its main function in the latter context.

Finally, rolling and cackling can become disconnected during the triumph call. This occurs if a gander is far away from his triumph-calling partner when he wins a fight and

Figure 96: *In the rolling call, the thickening of the neck through the lowering of the hyoid bone is clearly apparent.*

cannot reach the partner before his specific arousal evaporates. The specific arousal associated with the triumph call is apparently adapted for a standard distance between the partners.

Mamblona-Fischer regards rolling as a form of behavior that is largely motivated by conflict. This is true in part: aggression and the opposing bonding motivation are surely involved. But the inherent motivation of the unritualized behavior pattern is also involved. The latter must definitely be regarded as a demonstrative display, according to Heinroth's definition, which requires that it have a repellent effect on males and an attractive effect on females, as is often the case with so-called self-advertisement. A gander makes himself look as big and imposing as he can when uttering the rolling call (Figure 96).

Goose or Gander?

Heinroth thought it astonishing that a gander, lacking the special display feathers of a peacock or the conspicuously colored inflatable throat sac of a frigate bird, is nevertheless able to make a show of himself.

Even a superficial acquaintance with a goose pair reveals a striking difference in behavior between males and females. It is therefore hard to believe that the ethogram shows not a single innate motor pattern that is sex-specific, that is, confined to one sex.

The difference between the sexes lies in the frequency with which certain motor patterns are performed. We occasionally see the bent-neck posture performed by females, but we have observed it thousands of times with ganders. Conversely, we occasionally see ganders performing the so-called female tasks, such as the transfer of material during nest building or production of the hoarse call in response to small goslings. But we would not be able to tell a human observer how to distinguish males and females reliably, in every case, on the basis of a specific behavior pattern. It would appear that the neurological-anatomical basis of the behavior patterns, the "wiring," is the same in males and females but that there are major differences in hormonal conditions and in the physiological aspects of stimulation that govern the elicitation of different behavior patterns in the two sexes.

Sexual Recognition

We know of many vertebrate species, including mammals, birds, and fish, and many arthropod species in which there is no external difference between the sexes, and yet males and females recognize each other without hesitation. That is, there is virtually no attempt to form homosexual pairs. Sexual distinction is ensured by a mechanism that was discovered and analyzed by Beatrice Lorenz-Oehlert. It is important, however, that there be an adequate number of potential partners of both sexes. When the choice of partners is limited, two males or two females may form a pair, with the dominant individual adopting the male role.

In all animals, there is a remarkable ambivalence in sexual behavior. Each individual has at its disposal closed sets of male and female behavior patterns. Components of the two sets are usually not mixed, and which set actually functions depends solely on the relationship to the mate in the pair. Given freedom in their choice of a partner, females will select dominating partners, and males will select partners that prefer to be dominated. The appropriate one of the two sex-linked sets of behavior patterns will then operate. A bird with a dominant partner behaves as a female, and the dominant partner behaves as a male, regardless of their actual sexes. This is known to be so with domestic pigeons, as well as with jackdaws and probably corvids in general.

A series of experiments with pairs whose partners differed in strength has convinced us that with two unfamiliar individuals in full reproductive condition three autonomous kinds of arousal are simultaneously activated in encounters between conspecifics: first, an aggressive motivation; second, the motivation to mate; third, the escape motivation.

What differs between the sexes is the miscibility of the three kinds of arousal. In the middle of courtship, the male may physically attack the female; but in the female, sexual arousal is completely extinguished as soon as she becomes dominant over a male or female partner. Even if the subordinate partner is a male, the female is unable to respond to him sexually. This distinction between males and females appears to be widespread among higher vertebrates.

When two unfamiliar, sexually mature individuals of such a species encounter each other for the first time, a specific sequence of behavior patterns is performed. Both animals adopt a demonstrative posture, and each attempts to gain the upper hand in the social interaction that follows. If the two individuals are of the same sex and evenly matched in strength, a fight ensues. When the outcome is decided, one of them flees and usually will not challenge that opponent again. If one of the two individuals is a female, a transition from fighting behavior to sexual behavior occurs at some stage, often quite abruptly.

The parrot breeder Norbert Grasl once attempted to determine the sex of waxbills, which are indistinguishable externally, by arranging encounters between two individuals after a long period of isolation. Because mating occurred at once, he mistakenly thought he was able to distinguish the sexes. Then he discovered that the last bird placed in the test cage, still startled by the procedure, was consistently adopting the female role, with the other behaving as the male.

Such ambivalent behavior has been investigated in detail in the North American ruffed grouse (*Bonasa umbellus*) by A. A. Allen. The determination of male or female behavior proved to depend not on hormones but solely on stimuli emanating from the partner. We know of cases in which birds of this kind have been paired with two partners, one dominant and one subordinate, and have switched between the male and the female role in a matter of seconds. With such birds, homosexual pairs, either male or female, can be established at will. Often pigeon breeders do not know they have mistaken two females for a breeding pair until they find four eggs (instead of the two to be expected from a single female) in the nest.

So far, we know of no such partner-dependent determination of sexual behavior in anatids. While a hen without a male partner will crow and attempt to mount subordinate conspecifics, an isolated female greylag goose remains always a female. I have never seen mounting behavior between pure-blooded female greylag geese, although it occurs sometimes with domestic geese and with domestic-wild hybrids. The only case we know of stable pair bonding between two anatid females has been reported for the South American Chiloe wigeon (*Anas sibilatrix*). But this is an unusual species, in that the female's display plumage is exactly the same as the male's. When two male ducks form a pair, as can easily be contrived in an imprinting experiment, both retain clear-cut male behavior.

Trios consisting of a female attached to a pair of ganders are relatively common. By contrast, trios of two females bonded to a single male are rare, occurring only when the two females are bonded to the same male and are held together for this reason. In two of the cases I have observed personally, the two females were sisters from the same clutch, closely resembled each other, and were still attached to each other by close family bonds at the time of pair formation. In such cases, I am inclined to the interpretation—unbelievable to some—that the gander is unable to tell the two females apart if he cannot compare them side by side. On one occasion, the two females of one of these trios nested close together and could not prevent their broods of goslings from becoming inseparably intermingled. This was the context for one of the few wing-shoulder fights we have ever witnessed between female greylag geese.

It is another question whether bonding between two ganders indicates a failure of the Lorenz-Oehlert mechanism. A gander pair is similar to a heterosexual pair in so many ways that it seems likely that only one factor could lead to such a bond. Certainly the copulatory motivation plays no part.

Pair Formation

The unity of goose pairs constitutes the basic foundation, the infrastructure, of the social system. To a certain extent, an unpaired greylag goose is an incomplete representative of its species. As we have said, a pair may sometimes include two ganders but never two female geese. The process of pair formation usually begins in early spring, when the geese are maturing sexually. Because goose pairs are often broken up by the hazards of the natural environment, however, it would not be advantageous for the tendency to pair formation to be seasonally restricted. The abrupt onset of infatuation can strike a goose at any stage of its annual reproductive cycle. Even brooding geese or those that are about to begin brooding can be susceptible to courtship by an attractive stranger. Conditioning by means of rewarding stimuli is not necessary, and therefore there is a certain degree of similarity of infatuation to imprinting.

A friendly relationship that can prepare the way for later pair formation often develops when two individuals form an alliance through chance. It may be that a previous partner has died, or both geese have been sold into a strange flock and thus have become somewhat dependent on each other. With the rarer goose species, animal dealers often try to contrive pair formation by deliberately applying this method.

Demonstrative Behavior

In the first encounter—that is to say, during the mysterious, instantaneous process of infatuation—the demonstrative behavior of the two partners plays a major part. Following Heinroth's definition, we will take demonstrative behavior to include all the patterns that have the effect of both attracting a partner and driving away rivals. Because there are many patterns that serve these two functions, the concept is rather loosely defined. In all animals in which the male ranks higher than the female, as with most bony fish and birds, demonstrative behavior is primarily a property of the male.

The behavior often involves an excessive display of force, as William MacDougall observed in coining the term conspicuous waste. Nevertheless, the number of behavior patterns that are exclusively concerned with demonstrative display make it appropriate to include descriptions of them in our ethogram of the greylag goose.

Demonstrative behavior, even when not accompanied by an increased readiness to fight, is structured in such a way as to display the properties of the future father that possess survival value. A courting gander exhibits a remarkable shift in his thresholds for stimuli that will elicit an attack on a rival conspecific or else flight from a threat presented by another species. Just as his courage in the face of conspecifics is increased, all his thresholds for alarm calls and escape behavior are lowered. In a goose family, the leading gander is always the first to raise the alarm against an outside threat and the last to flee from a rival.

The Frigate Posture

Certain demonstrative displays of the greylag goose are specially ritualized to make the displaying bird look bigger. A displaying gander rides high on the water, his wing feathers ruffled and the wings raised from the shoulder joints so that the elbows stick up and in the side view display the entire wing. The rear end of the body is also raised high, reminding one of an old sailing ship, and, as we have said, that is why we call it the frigate posture (see page 188 and Figures 66, 67, and 68). For anyone familiar with anatids, the posture gives the impression of a display of the gander's readiness to fight because of its similarity to the threat posture of the mute swan. In Heinroth's definition of demonstrative behavior, it simultaneously indicates readiness to fight and courtship. But

Figures 97 and 98: *Demonstrative vigilance. A sentinel gander will not allow any other gander to display this behavior nearby.*

I classify the frigate posture of the greylag goose clearly under the heading of courtship behavior, for a gander showing this posture will never proceed to attack.

Demonstrative Vigilance

Vigilance, especially demonstrative vigilance, is particularly in evidence when a flock of geese is grazing in unfamiliar territory. Vigilance is shown by specific individuals, for which we use the term sentinel ganders. These ganders do not keep watch with the beak horizontal, as is otherwise the case, but adopt a posture that is influenced by demonstrative behavior and that we therefore call demonstrative vigilance. The bird stands with its body extended upward, the shoulders expanded and the chest puffed out. The beak is angled upward, and the eyes bulge partway out of their sockets as a result of the arousal of the sympathetic nervous system (Figures 97 and 98). At the highest intensity of specific arousal, the gander sometimes performs peculiar sideways twisting movements with the head and neck, which undoubtedly signify a threat of some kind.

Every sentinel gander treats demonstrative vigilance as his

special privilege. He will prevent any nearby gander from showing such vigilance, immediately attacking if an attempt is made to do so. The only bird a sentinel gander will allow to show demonstrative vigilance alongside him is his own mate, although female geese show such behavior infrequently. The ganders engaged in vigilance studiously avoid each other, often taking up positions on the periphery of the flock. These positions are often close to cover and thus are nearest the spots from which danger is likely to threaten.

Because of this separation, high-ranking sentinel ganders have little contact with each other. If the flock becomes crowded together—for example, by the scattering of high-quality seed during feeding—it can happen that two sentinel ganders will draw close unawares. When alerted, they may raise their heads simultaneously and find themselves looking into each other's eyes, leading to the remarkable phenomenon that M. R. Chance has termed cut-off behavior. The two ganders pretend not to have spotted each other, angle their gaze slightly away, and move off casually in opposite directions.

The frequency of demonstrative vigilance varies greatly from one individual to another. Some ganders never attain the rank, and others have personal experiences that result in a modified reinforcement of demonstrative vigilance. The gander Airotsohn, who hatched in 1970, took over the role of partner to his widowed mother while still quite young, and was so assiduous in this role that he tended to show an exceptionally strong demonstrative vigilance. Later, with his own mate, he bred outside the colony and spent only the autumn and winter in Grünau. Yet even today he excels over all the other ganders in his tendency to perform demonstrative vigilance—and it was this that enabled me to recognize him when he returned in July 1985 after a lengthy absence.

Demonstrative vigilance is particularly conspicuous when the gander is in the early phase of courtship, while the female is still reticent and the gander struts alongside her matching his steps to hers, in the manner Heinroth has described.

Courtship with the Sneaky Posture

There is one particular stage in the life of a greylag goose when its self-confidence is reduced to the minimum. This is right after the young goose's family has dissolved and the individual has lost its original status in the hierarchy of the goose flock. Such low-ranking youngsters tend to adopt the submissive posture (see page 203). It is just at this seemingly inappropriate time that the first contacts between the sexes begin to take place, although we did not discover the connection until quite late in our studies.

It is usually the gander that takes the initiative in courting, turning his attention to a female or sometimes to another male. He begins to follow the subject of his courtship, and in doing so moves somewhat faster than is usual for a goose showing the sneaky posture. That is what makes the gander conspicuous. He tries to anticipate the moves of his prospective partner by reading her intentions from her face, and this leads to the typical parallel walking shown by young pairs.

During this form of slow pacing, one of the birds will often point its beak sharply downward as if looking for food. When the movement is performed from the sneaky posture, there is a brief and coincidental resemblance to the typical bent-neck posture. But the motor impulses of the sneaky posture and the downward glancing seem to gradually become attached to each other, and this eventually yields the ritualized bent-neck pattern. A downward-directed tendency in this behavior pattern that is independent of food seeking is apparent from the fact that the courting bird will often bite at the rings on its leg. The submissive attitude of the male is particularly conspicuous, and it progressively leads to the suppression of the female's escape responses, to the point that she accepts the gander's presence directly alongside. At this stage, even the cackling of the gander, during which he brings his head quite close to the female's, no longer elicits an escape response. This timid-looking form of courtship thus has the same end result as the obtrusive courtship of a fighting gander.

It is my impression that when a somewhat older gander is courting a shy female, she will pretend not to look at him, giving him only occasional brief glances without turning her head. These are glances "stolen" from the corner of the eye and do not correspond to the direction in which the head is turned. In all birds, the eyelid is firmly fused to the eyeball and moves with it, in contrast to mammals (including humans), whose eyeball moves under an immobile eyelid. In birds with a colored margin on the eyelid, this is what indicates the direction of the gaze, as Heinroth showed in his study of pigeons.

In the placid and somewhat subtle form of courtship involving the sneaky posture, such eye play may well have an important part, although it is difficult to say, with this type of approximation between the sexes, which of the two partners is the most active. The observer tends to assume that the actively approaching bird is a male, but at this stage it is easy to be mistaken. We have no observations as yet to indicate whether a female who has learned to tolerate the proximity of the male through courtship with the sneaky posture will later accept his stormy arrival during the triumph ceremony.

In any case, the parallel walking gives rise to a bond that is rather like our rubber band in that its attractive force increases with the distance separating the partners. Such walking out together, to use the colloquial term for an early phase of courtship in humans, is not found in all birds. Many of them, such as woodpeckers and night herons, are held together almost exclusively by the common bond to the nest site.

Contrary to my previous belief, courtship with the sneaky posture is not limited to maturing young geese. I once saw an old gander who had been a widower for twelve years courting a seven-year-old goose with the sneaky posture and occasional bent-neck displays—and this was in July, after the molt.

The Bent-Neck Posture

The rigidly ritualized bent-neck motor pattern, which we have alluded to a number of times, must now be examined in detail.

We first discovered the bent-neck posture as an autonomous instinctive behavior pattern when Rolf Ismer sent us two solitary adult greylag ganders some years ago. These two males, who were attacked from all sides but were sexually quite active, moved around showing the posture almost constantly. I initially misinterpreted it as a compromise resulting from a combination of the arched-neck posture and inhibited threat. A. Tipler-Schlager cast doubt on this interpretation early on, suspecting a connection between the bent-neck display and the sneaky posture. And M. Martys believed that the orientation of the beak almost vertically downward indicated a feeding-intention movement.

Detailed analysis of long sequences of video film has now revealed that the geese, while walking alongside each other in the sneaky posture, apparently find things on the ground to peck at. In the transition from the sneaky posture to pecking, the precise pattern of the bent-neck posture is shown for a fraction of a second. The film sequences repeatedly show how this posture, initially brief and transitory, becomes abruptly frozen for gradually longer periods. To the human eye, the posture seems tense and unnatural, so it is fairly certain that it has a signaling function (Figure 99). In this sense, the bent-neck posture serves as an invitation to form a firm and lasting bond. For many years we regarded the bent-neck posture as a sex-linked male behavior pattern. But, as with many other behavior patterns, it is also present in females, although its performance requires a very high level of arousal and it is therefore rarely seen.

Figure 99: *The bent-neck display results from the ritualization of downward glancing performed by a goose in the sneaky posture.*

The Commonest Form of Pair Formation

Abrupt infatuation leads to a whole series of behavior patterns, which in turn lead to definitive pair formation. Indeed, the demonstrative behavior of the male, that is, his display of courage and strength, is successful only if the courted goose eventually accepts his invitation to join in the triumph ceremony and ceases trying to flee.

Courtship with the sneaky posture is relatively inconspicuous and can apparently be omitted, so that the gander can immediately begin with his demonstrations of strength. He seems to change suddenly, from one day to the next, performing all movements involving the voluntary muscles with exaggerated force. On dry land, the gander swaggers along with extremely long strides, his body twisting from left to right to an abnormal degree. His body is held high, his chest is expanded, and his anterior air sac is somewhat inflated. At this stage, he will blindly attack other living organisms, not just conspecifics but creatures he has previously been afraid of, as we saw with Martin (page 25). By means of victories achieved through surprise and good fortune, he can considerably improve on his original rank.

At the beginning of a courtship consisting of genuine or sham attacks followed by cackling invitations, one-year-old ganders will be heard for the first time uttering the vocalizations that we call rolling. The sequence of rolling and pressed cackling in the ritualized form of the classic triumph ceremony appears for the first time in the gander's ontogeny at this stage of courtship.

The courted goose may initially show no sign of noticing the demonstrative behavior of the gander, but she is fully aware that the behavior is directed toward her. As Heinroth has concluded, this is doubtless due to communication through eye language. We have already talked about the emphasizing of eye-movement signals by the margins of the eyes, which are conspicuous during the mating season.

While performing his excessive displays of strength, the gander is also demonstrating caring behavior, a further desirable quality in a future father. He shows vigilance and utters more alarm calls than any other individual. Demonstrative vigilance is characteristic of this stage of courtship.

Bonding Between Ganders

Persistent bonding between two males is a phenomenon that has so far been found only among species of the goose group. Ganders may join together as closely as the partners in a heterosexual pair—indeed, more closely. Opinions differ widely regarding this kind of bonding, however. Some observers regard it as a misfunctioning of the neural basis for pair bonding; others see it as indicating a general relationship of a presexual or suprasexual nature that should be considered apart from any sexual context.

Bonding between ganders is similar to the marriage of a heterosexual pair in terms of the behavior patterns through which the partners are united. An inexperienced observer could watch a flock of geese for a considerable length of time before realizing that some of the firmly united pairs consist of two ganders. Nevertheless, it is misleading to think of such

Figure 100: *An attempt at copulation between two ganders.*

a bond as homosexual, since copulatory relations do not necessarily exist between the partners. Copulation, in fact, plays only a minor part in the ordinary bonding between male and female. The joint performance of the triumph ceremony, for example, is far more important.

There are, indeed, gander pairs in which one partner regularly mounts the other—it is anatomically possible—but this is not invariably the case (Figure 100). The fact that there is no fundamental difference between gander pairs that exhibit mounting and those that do not indicates that copulation has no special significance as a bonding factor. But, because specific ritualized behavior patterns that have no direct connection with copulation are exchanged exclusively between mated pairs and between males that are bonded, the latter type of bond cannot be regarded simply as a form of friendship.

We often find ordinary friendships between geese and between geese and humans. When one living being actively seeks out another (even if the latter belongs to a different species) and remains attached under the influence of my metaphorical rubber band, staying close and greeting after lengthy separations, I regard this as a friendship in the usual sense of the term. I know of geese that actively greet a human friend far more intensively after an absence of several months than after a brief separation and then (in a response that has a strong emotional effect on the human) settle down to rest close by. I regard such relationships as analogous to friendship.

The term homosexual is misleading because it gives the impression that certain ganders are more inclined to form bonds with members of the same sex than with those of the opposite sex. Yet it is just as difficult to predict future bonding relationships for the members of a gander pair as it is for the partners of a heterosexual pair. Ganders that have lost their mates after long, fertile marriages may subsequently bond with other ganders, often with partners in the same situation. This happens with about the same frequency as the converse, in which the original members of gander pairs will pair off with females after the deaths of their partners. For example, the gander Veit immediately "married" a female goose after the death of his male partner. Therefore we cannot interpret the "homosexuality" of a gander as a deviant tendency.

Differences Between Gander Bonding and Heterosexual Bonding

A major difference between the two kinds of bonding resides in the fact that gander bonding may occur between brothers, but we know of only a handful of cases of heterosexual bonding between siblings. One of these happened because the two siblings had been separated long enough during their first years for them to forget that they were related. Another case involved two members of a group of hand-reared geese. The close relationship to the human foster mother is naturally weakened at the time of fledging, but hand-reared young geese tend to remain closely associated in a sibling unit. Because female siblings—apart from a few exceptional cases that we have observed—do not respond to the courtship behavior of their brothers, they usually leave the sibling unit after pairing off with strange ganders. The remaining brothers continue to stick close together. As a result, they are able to assert themselves more easily in interactions with other members of the flock than if they did not have this social connection.

Most brother pairs are derived from hand-reared flocks. In addition, young ganders that pair off with other males show a clear preference for partners of approximately the same age. Almost all of the eighteen gander pairs we have identified in our colony have been composed of individuals of the same age. In experiments with ducks and geese, Fritz Schutz has been able to demonstrate the same kind of preference among individuals that were reared together.

Another important difference between heterosexual and male pairs is that with the latter the bond need not be restricted to two individuals. In addition to our famous "quartet," we know of other cases in which more than two ganders have been linked together by bonding behavior patterns, especially the triumph ceremony.

In fact, a bond between two ganders does not prevent them from responding sexually to females, although they do not necessarily bond with those females through triumph-calling. In 1979 and 1981, the ganders Veit and Rufus associated with a certain female, with whom they twice raised a brood of goslings. Nothing changed in the relationship between the ganders when the female was snatched by a fox. The ganders Max and Kopfschlitz also reared offspring with a female, and after Kopfschlitz died, Max showed no bonding to the surviving female. (Peter Scott has observed that families with two fathers are relatively frequent among free-living pink-footed geese and that such trios are particularly successful in rearing offspring because both fathers take defensive action against predator gyrfalcons.)

As a general rule, gander pairs are more excitable and make more noise than heterosexual pairs do. A pair of geese that begins to utter rolling calls while still high in the air before flying in to land often turns out to be a gander pair. Because both partners clearly exhibit male behavior, gander pairs are more aggressive than heterosexual pairs, and they often become embroiled in conflicts with other members of the flock. Nevertheless, we know of no case of a gander pair remaining steadily higher in rank than a heterosexual pair leading offspring.

The greater number of "affairs" that occur with the members of gander pairs makes them seem more like newly established heterosexual pairs than old established pairs. The partners of a gander pair remain close, and the vocalizations they exchange are generally characterized by a preference for rolling calls, although they often perform pressed cackling "cheek to cheek" as well. In our experience, the duration of gander pairs does not differ from that of heterosexual pairs. Our oldest surviving gander pair has in fact been together since 1976.

Robert Huber once conducted a quantitative analysis of the behavioral differences between heterosexual pairs and gander pairs. It emerged that each of the males in a gander pair showed vigilance only slightly more often than other ganders did. But the doubling effect resulted in the gander pairs performing this behavior pattern about 2.2 times more frequently than the others, on average. Outside the courtship period, the partners in gander pairs did not show markedly higher levels of demonstrative vigilance than the ganders in heterosexual pairs did. Characteristically, however, the frequency of demonstrative vigilance increased significantly during the courtship period with males of the heterosexual pairs, whereas the level remained the same with the gander pairs.

In summarizing his results, Huber stated: "The special quality of the behavior of gander pairs derives from the fact that the partners perform the individual behavior patterns at a particularly high intensity. They make themselves obvious within the flock because of their frequent, exaggerated vocalizations. Attacks against inadequate targets or, in extreme cases, sham attacks against imaginary opponents are often observed. This seems to be brought about by the conspicuously higher level of general arousal that is found with gander pairs in comparison to heterosexual pairs. Because the partners generally exhibit male behavior, situations in which the male and female play different roles—for example, copulation or performance of the triumph ceremony—can lead to temporary disruption of the homosexual bond." In such situations, the partners of a gander pair may be seen engaging in a vigorous wing-shoulder battle. But immediately after the fight they usually show particularly intensive bonding behavior.

Disruption of Pair Formation Resulting from Weak Bonding

Let us turn now to the pair-formation behavior of some geese that showed intensive behavior patterns but achieved only partial pair bonding or completely failed to establish a bond. If the behavior patterns we observed between these weakly bonding geese are straightforwardly related, the account leaves an impression of complete confusion. One can set out the sequence in which a given goose associated with others, displayed invitations to participate in a triumph ceremony, performed cackling, and engaged in fights ranging from the simple chasing away of another individual to dangerous aerial battles. And when all is told, the overriding impression is of a vast expenditure of energy with no demonstrable biological reward.

We call these geese the unbonded ones, though it would be more accurate to say that they were poorly bonded. They were usually to be found in a loose association at the edge of the flock. As far as I know, the weak bonds between them did not yield a single successful breeding pair. These poorly bonded geese produced only two fledged offspring, whereas during the same period the pair formed by Sinda and Blasius raised no fewer than ten healthy offspring.

Jealousy

As has already been pointed out, the greylag goose's peculiar process of "falling in love" in many ways resembles its human counterpart, and it can lead to a similar disruption of all previous bonds. A young bride or a firmly paired female, even one already on the nest, can suddenly become interested in another partner. Conversely, a male may abruptly switch his courtship from one goose to another. In such cases, the abandoned partner, whether the male or the female, has some quite specific, instinctively programmed behavior patterns at hand. Because of their readily recognizable form, we include these patterns here.

Male Jealousy

In my experience, the most violent battles between two ganders are those that result when a female responds to both of them and does not know which one to choose. Although I cannot prove this hypothesis, it is certainly true that feuds are especially common between equally matched males that are courting the same female.

In other cases, when the female is clearly taking an interest in another gander, her mate shows a conspicuous type of behavior known as mate guarding (Figure 101). A female who has fallen in love with a strange gander rarely shows the timidity that is typical of an unmated female. A female that is changing allegiance will openly run toward the gander of her choice, while her "husband" attempts to prevent her from doing so. We will see all three geese hurrying along in this order: in the lead is the female's chosen gander, not necessarily interested in her; next is the female, following behind, and after her, her mate, intervening and trying to block her path,

247

Figure 101: *A guarding gander bars his unfaithful mate from approaching his rival.*

uttering pressed cackling with his neck stretched far out in front. Sometimes the mate will even try to push the female away with his shoulder, aiming inhibited bites at her shoulder.

If the gander the female is pursuing is not interested, he will passively run away, holding his head high, his neck thickened by the maximal ruffling of the small feathers on his neck (see page 205 and Figure 81). This is particularly so if the gander is not free to respond to the female, for example when he is already bonded to another. The thick neck to some extent conveys the message "I would like to get away, but unfortunately I must stay where I am."

By contrast, when the strange gander reciprocates the love of a female being guarded by her mate, embittered fighting often breaks out between the two ganders, and the blows from their wing shoulders can be heard far and wide. If the female's original partner is beaten, that does not necessarily mean that he will give up all claim to his mate. We have observed a badly beaten gander still trying to guard his mate. Guarding behavior is also shown by ganders when they are attempting to keep their mates away from possibly dangerous situations. If a tame goose that is accustomed to greeting and approaching humans is mated with a shy gander, the gander's guarding behavior can be elicited at any time by simply inciting the goose to approach. Guarding behavior is also seen in gander pairs when one partner shows an inclination to switch allegiance.

If the infatuation of a goose with a strange gander is reciprocated and the latter then tries to come between the pair, the "husband" may display the sneaky posture toward his mate. This is particularly common when neither gander is strong enough to engage in a decisive fight. It can also occur when the female's mate is intimidated by the fighting prowess of the strange gander but the latter has not responded strongly to the female's courtship.

In the late autumn of 1986, I observed one peculiar form of jealous behavior with the gander Veit. He had lived for many years in a gander pair with his brother Rufus, until in November 1985 the latter was run over and killed by a car on an icy road. The "widower" showed little grief and soon formed a bond with a young female. A few months later, another female goose attached herself to the pair. From then on, the gander behaved jealously with both females. Whenever they tried to join in his pressed cackling, he would drive them apart by pushing between them and uttering roll-cackling calls to either side.

Female Jealousy

The female goose also possesses behavior patterns that she performs in order to prevent the gander from being unfaithful. When a mated gander falls in love with a strange female and presents her with a triumph-calling invitation, the female has a quite refined method of intervention. At the point when the gander has completed his sham attack and is about to utter pressed cackling toward his new love, the "wife" runs rapidly toward her rival, blocks the path of the approaching gander, and begins to cackle intensively. The ritualized coupling of rolling and cackling is rigid enough that it obliges the gander to keep to the behavioral sequence of the classic triumph ceremony. He therefore completes the ceremony with his regular partner, and she participates in his pressed cackling with great intensity. This diversion of the triumph ceremony strikes the observer as highly comical, for it is easy to see a parallel in human behavior.

Rivalry at the Nest Site

Extremely intense fighting responses are evoked from the ganders when two pairs happen to select the same nest site. In 1985, a fight over a nesting box broke out between the ganders Muck and Siegfried, with only minor participation from their mates. The nesting box stood alone on an island in what we called the trout pond and was perhaps especially attractive for this reason. The ganders repeatedly engaged in aerial battles, and it was enough for one of them to appear in the sky a long distance away for the other to take to the air in preparation for battle. Aerial battles also took place in a dispute over a nest site on the floating island in Lake Alm.

Jealousy Between Flocks of Goslings

Another, very intensive form of jealousy must definitely be regarded as an artifact. It arises when a human who is already leading one imprinted flock of greylag goslings tries to take over a second flock. We used to try this whenever fewer caretakers than flocks of goslings were available, but it proved impossible because of the intense enmity an older flock of goslings would show to a younger flock.

Apparently the situation is different with ducks. When two flocks of ducklings are combined, as is the common practice on farms, there is a brief period of embittered fighting. This soon leads to confusion, as ducklings from the same brood begin to fight each other, but then the fighting wanes rapidly.

Grief

A greylag goose that has lost its partner shows all the symptoms that John Bowlby has described in young human children in his famous book *Infant Grief*. The tonus of the sympathetic nervous system declines, and as a result the musculature becomes limp, the eyes sink deep into their sockets, and the individual has an overall drooping appearance, literally letting the head hang. If we say this about a fellow human, however, we are referring not to the body posture but to the underlying mental state.

Small goslings that have lost their parents do not grieve in silence but cry loudly. They are completely incapable of engaging in any other activity. They stop eating and drinking and simply wander around crying. If their crying is not quickly appeased, they can suffer serious damage. Under natural conditions, of course, such *perditos* have no chance of surviving unless they find their parents. Only rarely will they be able to join another family or find a substitute pair of parents. It is therefore understandable for the goslings to use every scrap of their energy in seeking what they have lost.

Young geese that lose their parents after fledging show extreme grief accompanied by all the symptoms Bowlby describes. Although their behavior is otherwise normal, they restlessly seek their lost parents and continually utter distance calls. The participation of young geese in the greeting cere-monies and particularly in the triumph ceremony of their parents appears to be extremely important for their psycho-logical and physical well-being. When a bereaved young goose spots an adult sleeping with its head tucked under one shoulder, it optimistically takes the masked individual for the missing parent and rushes up in greeting, only to retreat plaintively as soon as the adult shows its head and proves to be a stranger. In geese, the physiognomy of the head is recognizable through characteristics associated with the beak and around the eyes. Humans are much the same, as is shown when the face is masked.

Geese show the most intense form of grief over the loss of their triumph-calling partners, although the duration of the mourning varies greatly. Heinroth has reported cases of widowed geese that called after their missing mates for years, especially during the breeding season. In other cases, we have observed a widow or widower forming a bond with a new partner within a few days. In individuals that have been mourning for a long time, the reduced tonus of the sympathetic nervous system is noticeable from the expression of the eyes. My friend Erich Bäumer aptly observed, when he saw our female goose Ada for the first time when she was very old: "She must have been through a lot!"

When a goose loses its triumph-calling partner, even one that is quite old and had been mated for some time may seek to reestablish contact with its previous family, although it may have had no contact over the intervening years. Hand-reared geese will show a clear tendency to remain close to their former caretaker. Gudrun Lamprecht-Bracht and I had an unforgettable experience of this kind one time when we were at the moor in Seewiesen observing our flock's evening departure for Lake Ess. After the flock left, we noticed a greylag goose squeezed between us, pathetically frozen in an extreme form of the sneaky posture. Because we cannot recognize a goose from a "bird's-eye" level, we both crouched down to identify the bands. It was Max, the male who had been paired for several years with the gander Kopfschlitz. As we got up, we both said sadly, in one voice: "Kopfschlitz is dead." We were right.

The fact that grieving geese sometimes disappear suddenly is probably due to their increased susceptibility to accidents. It is also possible, however, that they simply decide to leave the area, especially if they have suffered several setbacks in a row. We do not have a record of a single instance in which one of these geese has ever reappeared.

The loss of the triumph-calling partner extinguishes any trace of aggression in the surviving bird. However high the rank of a bereaved gander may have been, he will now allow himself to be chased away by the weakest and lowest-ranking conspecifics, showing no sign of resistance. Because geese, like jackdaws, are particularly aggressive toward individuals that have once been dominant, these widowers lead an extremely pitiful existence on the periphery of the flock, exhibiting the sneaky posture most of the time. As a rule, the situation changes only if the gander forms a new triumph-calling relationship. A gander that has lost its mate will often form such a relationship with another gander.

The loss of a triumph-calling partner thus results in major changes in the psychological and physical condition of a greylag goose, while the loss of small goslings is accepted as a matter of course. The loss of offspring, even those that are several weeks old, usually fails to elicit intensive seeking behavior, except when a large part of the flock turns up missing. Individual relationships apparently do not form until the offspring are somewhat older, although I cannot pinpoint the exact age at which active seeking of the offspring will occur.

On the other hand, I once observed a female snow goose showing active seeking behavior when one of her three offspring died suddenly after the goslings were over four weeks old. The goose was running constantly to and fro and could easily have lost her remaining goslings.

In view of the intensity and, sometimes, the duration of the grief a greylag goose shows over the loss of a partner, I find it remarkable that a dog shows no marked signs of grief for another dog, though it may grieve for its master. In her work with chimpanzees, Jane Goodall has described the grief shown by a young male who was physically and nutritionally independent of his mother at the time of her death. Even though he was mothered by his sisters, his grief gradually turned into a neurosis, and he eventually died of it.

Hatred

Any definition of the word hate must encompass the fact that a specific individual is generally the target of this emotion. Although hate is expressed by means of aggressive behavior, it is not to be confused with the ordinary attack motivated by the aggressive drive. The characteristic feature of hate, in contrast to ordinary aggressive behavior, is its duration. An antagonism between two ganders can last for months or even years.

Personal hatred directed at a specific individual sometimes arises because two ganders are bound together in a persistent conflict situation from which they attempt to escape by means of violent aggression. (We know of no such cases with female geese.) A typical hate relationship arose between the ganders Markus and Blasius, whose battles could easily have been fatal. The unusual situation is perhaps attributable to the fact that they were courting three sisters that were bonded to their foster mother in a very tight relationship.

While this form of hatred arises out of persistent competition over a specific object, be it a nest site or a female, other forms of hatred can stem directly from an existing bond. It is well known in psychoanalysis that hate and love are closely linked, and some of our records show that certain ganders once bound together in a loving relationship have ended up hating each other. The most impressive case of this kind involved two male snow geese, whose general behavior is much like that of greylag geese. These two males split up after a drawn-out and violent duel, and at first they showed chasing behavior. The manner in which they avoided each other was, however, extremely interesting. When we confined them together in a holding cage, they would not look at each other. Instead, in typical cut-off behavior, they showed obvious gaze avoidance and performed orgies of displacement activities involving preening and bathing on dry land. They also engaged in occasional duels at long intervals.

Disputes often occur between paired ganders at the climax of the triumph ceremony. The sideways turning of the neck decreases until the birds are facing each other eye to eye, the cackling calls become coarser, and within an instant the two ganders may have seized each other by the shoulders and be flailing out violently with their wing shoulders. The mechanism underlying such disputes probably involves a process that was recognized by Jürgen Nicolai. When ritualized movements exceed a certain intensity, the ritualization effect is gradually lost. To put it another way, the original motor patterns emerge without the deactivating influence of ritualization. The bullfinch shows a ritualized form of dueling with the beak, which serves as a pure mating ceremony. Nicolai was able to demonstrate that at maximum intensities, which were experimentally contrived through lengthy separation of the partners, the ceremony gave way to serious fights in which the female was always the loser.

Paired ganders will sometimes engage in violent fights with each other, but such an outbreak of hostilities is not necessarily irrevocable, and often it terminates in aroused but friendly triumph-calling. In some cases, however, the hostility may last a lifetime, as was recorded with the two ganders Max and Odysseus, who separated permanently after a duel.

Hatred also leads to violent chases, and it is common to see the interesting phenomenon of "embarrassed" avoidance of encounters. A good indicator of the depth of hatred is the distance one gander will fly in order to attack another.

The bond between two ganders promotes the general capacity to show hatred, even though there seems to be nothing but untroubled "love" between the two. A bonded pair of brothers, Veit and Rufus, one day began to display open hatred toward my assistant Paul Winkler. Perhaps some incident during their sensitive juvenile period had led to their

Figure 102: *Two socially bonded ganders launching a joint attack on a hated human being.*

enmity toward him. They not only attacked his person (Figure 102) but extended their hatred to the institute's car, which he often drove. He was eventually forced to fend off their attacks energetically, and the ganders then turned to attacking the car—a typical instance of the cycling response identified by B. Grzimek.

One day, through no fault of the driver, Rufus was run over by a car on an icy road. Veit immediately paired off with a widowed goose, showing no obvious signs of grief. In late autumn of the following year, he displayed triumph-calling toward another female, just a year old. Following this double alliance, his long-standing hatred for Paul Winkler seemed to increase still more. He would take off for an attack whenever the institute car came into sight around a bend some 200 meters away and would try to get at the driver through the window. If Paul threw his jacket on the ground, Veit would attack that until he was completely exhausted.

Analogy

Human Response to Analogy

I turn now to the special cognitive value of the remarkably extensive analogies we can draw between the behavior of the greylag goose and many details of human social behavior. More than seventy years ago, Oskar Heinroth wrote the following: "In this paper, I have particularly emphasized aspects of communicative behavior. It has emerged that, insofar as social birds are concerned, these aspects are surprisingly reminiscent of human behavior. This is particularly true when the family—father, mother, and offspring—forms a long-lasting unit, as is the case with geese. The descendants of the sauropsids have in this respect developed emotions, behavior patterns, and motivations similar to those which we find in ourselves and consider human. Study of the ethology of higher animals—unfortunately, a field in which relatively little has yet been done—will increasingly bring us to the realization that our own behavior toward family members and strangers is based on motivational drives."

It is a good strategy to select for study those animals that exhibit a wide variety of humanlike behavior patterns—especially since we tend to become emotionally attracted to such animals. My father, who in his old age spent many hours in our garden in the company of my flock of geese, viewed their behavior in a completely naive and anthropomorphic light. He laughed at their demonstrative behavior, particularly when it misfired, became annoyed over the subordination of weak individuals, and loved the ganders leading offspring because of their courageous interventions on behalf of their families. He was familiar with the rank order of the geese and, with no scientific background, understood the birds amazingly well. My father, a great dog lover, coined one phrase that we often repeat: "Next to the domestic dog, the most suitable animal for association with human beings is the greylag goose."

Intuition and Natural Science

Our intuitive understanding of a higher animal is governed by human emotion—and probably by the emotions of the animals themselves—and is therefore quite different from natural science. But both aspects, intuitive understanding and scientific knowledge, are needed to give us the ability to make predictions. My friend Frank Fremont Smith has defined science as an undertaking "that renders things predictable." His definition applies to the intuitive as well as to the scientific comprehension of the world. If I open a window in the room of a house in which the thermostat is located, my prediction that the rest of the house will become overheated will surely be fulfilled. If, on the other hand, I predict that a friend will be pleased with my gift, this prediction has approximately the same probability of being correct, although its origin may be purely intuitive.

The magnificent collective undertaking of humanity aimed at reaching an objective understanding of the world and at creating an overall blueprint free of contradiction is no more than a few hundred years old. Our intuitive, unreasoning perception of the environment, and of the organisms that inhabit it, is incomparably older, indeed primeval. Science is superior to this primeval understanding in that it can trace complex phenomena back to their smaller, more basic, and simpler elements. This is achieved in an objective, logical fashion that anyone can follow, and the findings are incontrovertible.

The relatively simple and cohesive picture of the world that has resulted has given humanity an astounding degree of power over its environment, but many questions that affect mankind and the whole of the natural world have been neglected. The science of physics generalizes many widely understandable and all-pervading laws, but it neglects the structures that are used to abstract these laws.

An explanation of the world requires not only knowledge of natural laws but also an understanding of the specific structures of matter in which these laws operate. Newton's laws, for example, manifest themselves quite differently in the movement of a pendulum than they do in the rotation of the heavenly bodies in our solar system. Reductionism, the assumption that a complex system is really nothing more than a simpler one, is erroneous, for it neglects structure. The widespread belief that regression to increasingly simpler and smaller elements can be pursued ad infinitum is mistaken. To put it another way, it is wrong to believe that science consists only of reduction and can do without the description of structure.

Nor can natural science be pursued to the exclusion of human emotion, in the belief that it is possible to be objective by ignoring one's own feelings. Objective investigation has always required that we take our subjective feelings into account and include them in the overall cognitive framework. One example that I have often used in lectures is this: A child comes inside from the garden. I touch the child's cheek and notice that it is feverishly hot. But I know that my hand, which has just been immersed in cold water, is colder than usual and therefore perceives heat more strongly. For this reason, I do not conclude that the child is ill. My subjective perception has been rendered objective as a result of my understanding of its physiological basis.

The two alternative routes to knowledge of the world that are available to us as humans are so different from each other that many authorities regard them as incompatible. Herbert Pietschmann speaks of two roads, the intuitive one, which distinguishes between subjective truth and falsehood, and the scientific one, which distinguishes the correct from the incorrect. C. P. Snow has referred to two separate human cultures, the incompatible worlds of art and science. Paul Weiss caustically observed, in a conversation we had around 1978, that he was still able to use his "binocular" approach to see human beings as intact, three-dimensional entities. Finally, no less an authority than Max Planck, in a short communication published in *Naturwissenschaften*, argued that the methods of reasoning and cognition used in the natural sciences are not fundamentally different from those used by all humans in the course of their everyday contact with nature.

Proponents of the evolutionary theory of knowledge should always be aware when they are using one route in the course of their work and when they are using the other. They cannot avoid using both. It would be tantamount to a refusal of knowledge—the greatest sin against the spirit of research—to close one eye deliberately, in the sense of Snow, Pietschmann, and others. In the course of our investigations, we biologists are confronted with some systems for which the scientific approach seems the promising route and others for which the intuitive approach is more immediately productive. The former is most likely to apply with lower organisms, the latter with complex organisms. Between the two extremes, however, are a multitude of organisms that evoke our emotional participation yet at the same time almost cry out for objective analysis, at least for certain aspects of their behavior. It is of the utmost importance for researchers to know when to rely on their intuition and when objective analysis is more appropriate.

If we feel ourselves emotionally affected by the behavior of an animal, it is a clear indication that we have intuitively discovered a similarity between its behavior and human behavior. We should not conceal this in our description. Such similarity can have only two possible sources. It can occur either as a result of homology, that is, by the inheritance of the shared characteristic from a common ancestor, or as a result of analogy, due to convergent evolution under the influence of similar selection pressures.

We use the same word for the eye of an octopus as for the eye of a vertebrate, and when we use the word we do not feel that we must always add a qualifier to the effect that the octopus eye is not the same as the vertebrate eye. Organs of this kind exhibit a high degree of analogy, often extending to small details. It would seem that only a limited number of technical solutions are available to living organisms. The solutions we find are so similar that it is hard for the observer to believe that two such structures do not have some common blueprint. Only detailed comparisons of the anatomy and especially of the embryonic development of similar organs can provide convincing evidence that they have different origins.

Analogy as a Source of Understanding and Misunderstanding

Animal behavior patterns that are analogous to our own are perceived as related, and we find them appealing. At one time, the extrapolation of human behavior into the animal world was self-understood and generally considered valid. One only needs to look at the writings of Alfred E. Brehm, who attributed many human qualities and abilities to birds and mammals. Anthropomorphism has since acquired such a bad reputation in scientific circles that many ethologists avoid so much as a mention of similarities between human and animal behavior. But remarkable similarities in fact exist between human and animal behavioral systems—such as the struggle for higher rank, jealousy, and bonding behavior—and they require some kind of explanation.

It is understandable—although mistaken, from the point of view of the theory of knowledge—when a researcher concludes that the body-soul problem is insoluble and that therefore it is best to avoid trying to investigate the subjective experience of human beings. Such was the conclusion drawn by the behaviorists. For animal behavior, which is not hard to investigate objectively but is subjectively inaccessible, the conclusion seems reasonable. The philosopher Descartes stated it this way with regard to animals: *"Animal non agit, agitur."* It was Karl Bühler who first won scientific recognition for subjective evidence. And such great thinkers as Kant and Schopenhauer, who were not naive realists, never doubted the existence of "fellow human beings." Yet they could not have known of their existence except by virtue of the very sense organs they so despised.

For a thinker who accepts evolution, the evidence of subjectivity in both his fellow human beings and the higher animals is undeniable. This acceptance is in fact enshrined in animal protection laws throughout the world. Obliged to recognize the existence of subjective feelings in higher animals, we accept the ethical consequences.

Nevertheless, such recognition should not mislead us into thinking that we can fathom or replicate the subjective states of animals. Our feelings are simply an indicator of convergent adaptation. Similarities point to important research goals which may be only indirectly accessible: to know the conditions that govern our emotional responses as well as the relationships between animals.

The extensive and quantifiable analogies that exist between the behavioral systems of greylag geese and those of humans strongly argue that they have arisen through phylogenetic convergence under the influence of similar selection pressures.

These pressures may have operated in the past, or they may still be operating today. We cannot say what selection pressures are involved. We do not know whether jealousy, aggression, or the struggle for rank have a positive selective value for humans. But—and this is the point—we can conduct both quantitative observation and experimentation with our animal subjects. With our geese, we are in the fortunate position of being able to study certain behavioral systems and their functions over many generations. We can estimate their survival value by counting the numbers of offspring reared to adulthood. The value of a longitudinal investigation of a social behavioral system in a continuously observed and documented population increases exponentially with the duration of the observations.

The functions of aggression between rivals or jealous behavior seem self-evident in the greylag geese. On closer examination, however, it is not clear how to make the calculation of costs and benefits required by the sociobiologists. It would seem advantageous for the species when a plucky gander manages to win a suitable nest site and rearing area for his offspring, occasionally saving a gosling from a predator in the process. On the other hand, the continual friction and the demonstrable dangers that arise from fights between rivals have enormous energy costs. Ganders that are continually involved in conflicts are especially vulnerable to predators.

We do not deny—indeed, as objective observers of behavior, we dare not deny—that we are thrilled when an old, familiar greylag goose greets us joyfully on its return after a lengthy absence. The reality that we are investigating is the interaction between ourselves and the environment, between intuition and objective knowledge. "The process of knowing and the object of knowledge cannot be legitimately separated," P. W. Bridgman said. What we should not forget is that we cannot know, and probably will never know, what the goose itself feels. We can assume that similar processes take place in humans and in animals. Because these analogous structures concern us as knowledge-seeking human beings, we should regard it as a duty to investigate these processes as far as we can take them with the only means at our disposal, the scientific method.

It is my belief that the behavior of the greylag goose, because of its various similarities to that of human beings, is a particularly suitable subject for scientific investigation. I flatter myself that I can confine my tendency to anthropomorphize within narrow limits. On the other hand, I have not fooled myself into believing that a brilliant flash of inspiration directed my attention to this promising subject of research. The credit should go largely to the poetic insight of a Swedish schoolmistress, who translated the meaning of the summoning call of wild geese, emotionally but scientifically quite correctly, as: "Here am I—where are you?"

Illustration Credits

Plates (following page 126)

Hermann Kacher: XVI, 1 and 2. Konrad Lorenz: I. Michael Martys: II, 1 and 3; III, 2; IV, 1 and 3; VI, 2–4; VII, 2; VIII, 1 and 2; X, 1; XIV, 2. Anna-Maria Schatzl: XIII, 1–3. Angelika Tipler: II, 2; III, 1; IV, 2; V, 1 and 2; VI, 1; VII, 1; IX, 1– 4; X, 2 and 3; XI, 1 and 2; XII, 1 and 2; XIII, 4; XIV, 1; XV, 1 and 2.

Figures

Hermann Kacher: 15–17; 30, 31, 33, 48, 79, 81, 83, 84, 89, 91–93. Konrad Lorenz: 1 (watercolor), 14 (sketch), 19–22 from *Beobachtetes über das Fliegen der Vögel*. Michael Martys: 5, 12, 26, 27, 29, 34, 37, 40–44, 46, 50, 56, 62, 64–74, 78, 80, 87, 97, 98, 100. Anna-Maria Schatzl: 13, 57, 59. Angelika Tipler: 9–11, 18, 23–25, 28, 32, 35, 36, 38, 39, 45, 47, 49, 51–55, 58, 60, 61, 63, 75–77, 82, 86, 88, 94–96, 99, 101, 102.

2, 4, 6, the property of Konrad Lorenz. 3, the picture archives of A. Festetics. 7, 90, from Helga Fischer, *Das Triumphgeschrei der Graugans* (drawings by Hermann Kacher). 8, from Julian Huxley, *The Courtship Habits of the Great Crested Grebe*. 85, from Angelika Schlager, *Rangordnung zwischen Graugansfamilien*.

Index

receptor side of, 88
Ismer, Rolf, 56, 239

jealousy, 247–51
  between flocks of goslings,
    250–51
  female, 249–50
  male, 247–49
Jennings, H. S., 85, 89

Kant, Emmanuel, 262
Kalas-Schäfer, Sybille, 31, 39, 49,
    50, 211–13, 215
Kaspar Hauser experiments,
    100–101
Kawai, Masao, 11
Kawamura, S., 11
Kirchshofer, Rosl, 196, 207
Klopfer, Peter, 100
Knoll, Fritz, 9
Koehler, Otto, 9
Kummer, Hans, 10

Lagerlöff, Selma, 1
lamentation call, 39, 165–66
Lamprecht-Bracht, Gudrun, 252
Lang, Ernst M., 139
learning, 90–91
locomotion, 109–36
  slowed, 186
Lorenz-Oehlert, Beatrice, 230, 232
lost calls, 4, 15

MacDougall, William, 234
Mamblona-Fischer, Helga, 28, 100,
    162, 218, 219, 222, 224,
    225, 228
Martys, M., 239
mating, 189, 191
  see also pair formation
Matthaei, Ruprecht, 9, 12

maximum value precedence, 92
Mergler, Myra A., 136
migration, 41
mobbing, 172–73
molting, 33, 39–40
motivational analysis, 93–95

neck arching, 188–90
neck dipping, 188–90
nest-calling, 182–83
nesting, 26, 33, 60–62, 63, 65, 76,
    129, 177–79
  rivalry in, 250
nibbling, 156–57, 158
Nicolai, Jürgen, 75, 255

operant conditioning, 90

Packard, Jane, 213
pair formation, 25, 28, 42, 44, 47,
    50, 71–73, 75, 194, 233–46
  commonest form of, 240–41
  disruption of, 246
palaver, 2, 39
parallel swimming, 62
parallel walking, 25
path conditioning, 22
Pavlov, Ivan, 90
pecking, 153–54
pecking order, 206
  see also rank order
pestering call, 166–67
phenocopy formation, 97
Pietschmann, Herbert, 260
Planck, Max, 260
plucking, 155
point landing, 129–30
polite-alarm display, 72
Portielje, A. F. J., 86
postcopulatory display, 189, 192
predators, 39, 42